CW00405473

3-99

REED'S SCHOOL

FIDE

Presented by _D. Thompson_

Date _June 2005_

Springer

London
Berlin
Heidelberg
New York
Barcelona
Hong Kong
Milan
Paris
Santa Clara
Singapore
Tokyo

Patrick Moore and Allan Chapman

Patrick Moore's Millennium Yearbook

The View from AD 1001

With 31 figures

Springer

Illustrations bearing the initials A.C. or R.E.W.C. were
prepared by Allan Chapman or Rachel Chapman respectively

ISBN 1-85233-619-6 Springer-Verlag London Berlin Heidelberg

British Library Cataloguing in Publication Data
Moore, Patrick, 1923-
   Patrick Moore's millennium yearbook: the view from AD 1001
   1. Astronomy, Medieval   2. Astronomy, Arab   3. Constellations -
   History
   I. Title   II. Chapman, Allan, 1946-, Alan
   520.9'3
ISBN 1852336196

Library of Congress Cataloging-in-Publication Data
Moore, Patrick.
   Patrick Moore's millennium yearbook : the view from AD 1001/
Patrick Moore and Allan Chapman.
      p.   cm.
   ISBN 1-85233-619-6 (alk. paper)
   1. Astronomy, Ancient.   I. Chapman, Allan, 1946-   .   II. Title.
III. Title: Millennium yearbook.
QB16.M67   1999   99-37448
520'.9'021—dc21   CIP

Typeset by EXPO Holdings, Malaysia
Printed and bound at the University Press, Cambridge
58/3830-543210 Printed on acid-free paper SPIN 10729892

This Yearbook is respectfully dedicated to
HIS MAJESTY ETHELRED II
King of Wessex and all of England

Contents

Editorial

It is necessary to apologise for the slight delay in issuing this Yearbook. The main text was in fact completed well on schedule, but most unfortunately one of the Danish raids, with which we have become so distressingly familiar, resulted in the destruction of all the copies which had been sent from the Winchester Scriptorium, and these had to be recopied elsewhere. Our armed forces, under King Ethelred II, are doing all that is possible to maintain order and protect our coastline, but it cannot be denied that at times the situation becomes very difficult.

◆ One very important change has been made for this new Yearbook. Up to now we have adhered to the time-honoured numerical system devised by the Romans, but it is true that the system does have its disadvantages; we appreciate that, for instance, multiplying XLCVIII by LXXXIV takes time. In the year 595, according to Christian reckoning, Indian mathematicians introduced a new concept, the zero (o), and from 815 many countries have used a totally new mathematical notation based on that of the Arabs. It is this which, after lengthy discussions, we have decided to adopt, so that in the present book the new system has been followed throughout. A conversion table, for those who prefer the old system, is given on page xi.

◆　This first edition has been produced in Anglo-Saxon rather than Latin, which will, it is felt, increase the readership. A Latin version will be prepared later if there is sufficient demand. Finally, it has been decided to separate the astrological sections from the purely astronomical text, and these sections will be issued in the near future.

◆　We extend our grateful thanks to the compilers of the Anglo-Saxon Chronicle for permission to reproduce data and diagrams.

Winchester
22 November 999

Conversion Table

New	Old
0	(no equivalent)
1	I
5	V
10	X
50	L
100	C
500	D
1,000	M

In the unlikely event that readers need reminding about the "old" method, in general Roman symbols are used in order of decreasing size: thus XVI = 16. A symbol may be repeated up to three times (XVIII = 18), but not four; thus 19 = XIX, not XVIIII.

The New Millennium and the Christian Calendar

This Yearbook is for the year 1000. It is thus, the last one for the current Millennium. But, you may ask, when does the New Millennium begin, and when does the old one end?

◆ Let us be clear. Our ancestors, even the illustrious Greeks, had no symbol for zero. Therefore there was no year 0. What we call "1 BC" was followed by "AD 1". This means that the first day of the new Millennium was the first day of AD 1; and from this, it is apparent that the first day of the coming Millennium will be the first day of 1001. Nothing could be clearer.

◆ There are some ignorant fellows who will continue to protest that the first day should be 1 January 1000 – just as, no doubt, equally ignorant fellows will protest that the next Millennium will start on 1 January 2000. But in this Yearbook we are better-informed.

◆ Let us, then, look forward to our new era, starting, by God's will and the laws of science, on 1 January AD 1001.

The Editors

The Night Sky in 1000

◆   The charts on the following pages have been drawn for mid-evening in Britain for the four seasons of the year: Winter, Spring, Summer and Autumn. We have depicted constellations mentioned by Ptolemæus (Ptolemy), who listed 48 in all, and have named them in the Latin form, as is generally accepted. The names of the brightest stars – mainly Arabic – have also been included.

◆   The charts are drawn for the latitude of our capital city, Winchester. There are, of course, stars further south in the sky which are never seen from here. Such is the star known to Ptolemy as Canopus, in the constellation of Argo (the Ship). This star can never be seen from Athens, but it can appear briefly over Alexandria, where Ptolemy spent most of his life. According to descriptions sent back by marine navigators it is almost as brilliant as Sirius. No doubt there are stars much further south in the sky which have never been studied, and about which no information is available.

◆   The positions of the planets for each season are included, but it is difficult to show Mercury, which moves so quickly. Remember, both Mercury and Venus move round the Earth in orbits which are smaller than that of the Sun, and they are therefore the closest of all heavenly bodies apart from the Moon. About the planets themselves we know little, though it is reasonable to suggest that they are not very unlike the Earth, and may even support life.

Winter

It is surely right to begin our yearly tour of the sky in winter, because our skies are then at their darkest and the stars shine down from the firmament with all their brilliance. The Moon, of course, does conceal all but the brightest stars when full; this year the Moon is full on 23 January, 22 February and 22 March. Then the watcher can see the Moon's glowing lands, and also the dark seas in which the inhabitants no doubt disport themselves.

◆   Of the planets, Venus begins the year in the evening sky, but sets very soon after the Sun. It will remain brilliant all through the Winter, though during the Summer it will be lost from our evening sky. No doubt this planet is a wondrous world; what beings might live there, and how do they see our Earth? It is right to name Venus after the mythological Goddess of Beauty.

◆   Mars, the Red Planet, so aptly named after the god of war, is high on the sky, close to the heavenly Twins, Castor and Pollux; it is now at its brightest for the year 1000. It will remain in Gemini throughout the Spring, then passing first into Cancer, the Crab, and thence into the Lion, fading as it does so; from outranking all but the leading stars, it will drop steadily until it is no better than the second

magnitude. Mars will be very near the Moon on
22 January, 18 February, and 17 March, while Venus
and the Moon will be close on 11 February and
12 March.

◆   Of the remaining planets, Jupiter, among the
stars of the Archer, is not visible; elusive Mercury may
be caught scuttling across the morning sky in February
and March.

◆   The scene is dominated by Orion, the celestial
Hunter, with his two striking leaders, the orange-red
Betelgeux and glittering Rigel. Note too the three stars
of the Hunter's Belt, which upward show the way to
Aldebaran, the orange Eye of the Bull, and downward
to the Dog Star, Sirius, the brightest star in the entire
sky. Sirius is white; so says Al-Sufi, and who can
disagree? Yet some of the older observers called it red.
Even Ptolemy did so. There must be some mistake
here: stars cannot change their hues.

◆   Note too the Pleiades, known since before the
dawn of history. Homer, in the Odyssey, tells us how
Ulysses "sat at the helm and never slept, keeping his
eyes upon the Pleiades …" and in the Bible, Job asks,
"Canst thou bind the sweet influences of the Pleiades
or loose the bands of Orion?" Al-Sufi lists four stars
in the Pleiades; other accounts give us as many as
seven.

◆   In the west, the Flying Horse, Pegasus, is sinking
into the twilight; in the east, the Lion makes his
triumphant entry. And, of course, the Great Bear,
Ursa Major, is always with us. It is in the far north of
the sky; Aristotle, the Greek, said that it was so named
because only the Bear is courageous enough to brave
the frozen regions towards our own earthly pole.

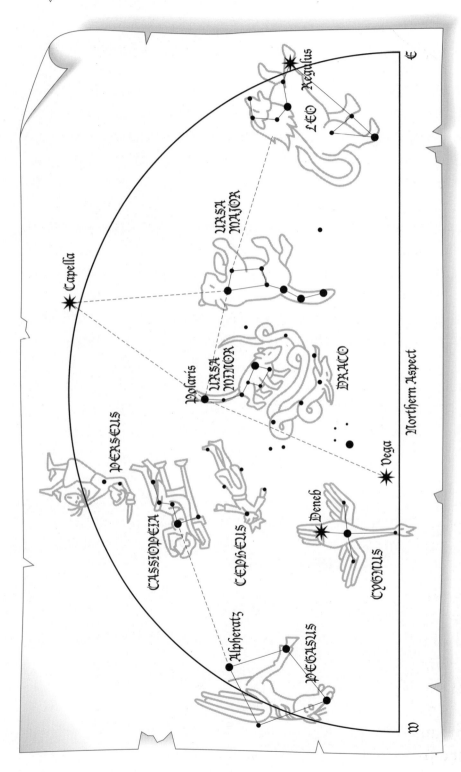

The stars of Winter, as seen from Winchester (latitude 50°) at 10 p.m. on 15 January 1000.

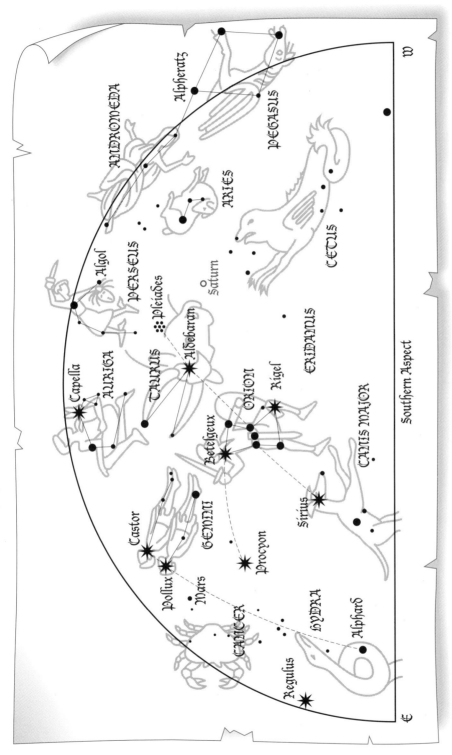

The stars of Winter, as seen from Winchester (latitude 50°) at 10 p.m. on 15 January 1000.

Spring

With the coming of Spring, the snows melt and the days lengthen; let us pray that in our year of grace 1000 the arrival of the warmer weather does not bring with it a new incursion from the Danish heathen. Mars remains in Gemini, still close to the Twins, but is losing his lustre, and by Summer he will have moved into Leo. Mercury, too, is almost absent, though Venus remains to adorn the evening sky for a while before sinking into the evening twilight. Jupiter may perchance be glimpsed before the Sun rises, but is very low down. We must wait patiently for him to rise higher. Saturn is now about to leave the evening sky, but never fear – the slow-moving planet will be with us once more before 1000 comes to its end.

◆   The Moon is full on 21 April, 21 May and 19 June. This decrees that for some nights around the "longest day", 16 June, there will be no true darkness over our islands.

◆   Cast your eyes toward Leo, the Lion of the sky; the curved line we sometimes call the Sickle is led by the Royal Star, Regulus. By June, Mars will be approaching Regulus, and will be no brighter than it – very different from the glaring Red Planet of January. And near Mars in late Spring make sure you look for the curious little patch in the constellation of Cancer, the Crab. It seems to be a veritable cluster of little stars; it was called by Hipparchus a "cloudy star", and has been likened to a swarm of bees. Others know it as Præsepe. The Greeks tell us that if the cluster cannot be seen in an apparently clear sky, rain is on the way.

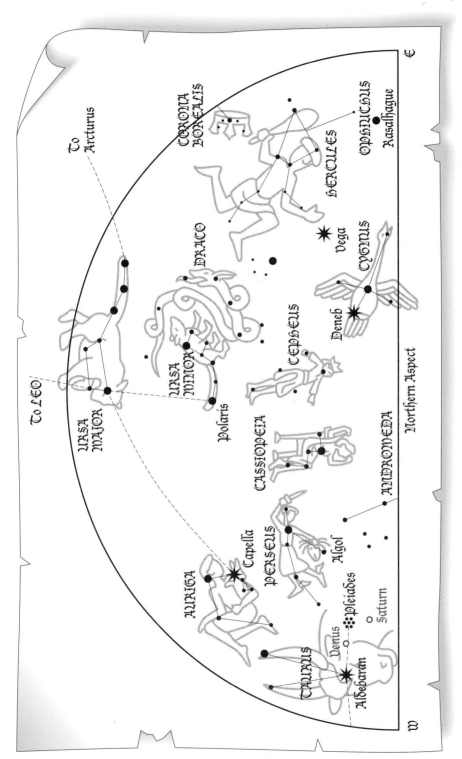

The stars of Spring, as seen from Winchester (latitude 50°) at 10 p.m. on 15 April 1000.

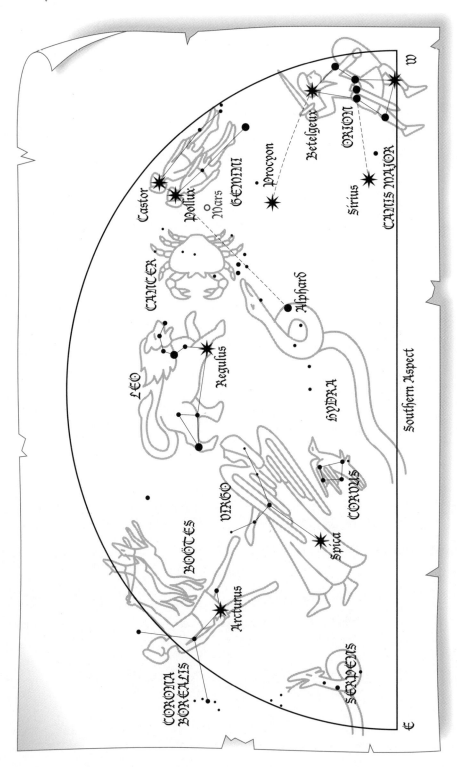

The stars of Spring, as seen from Winchester (latitude 50°) at 10 p.m. on 15 April 1000.

◆   Seek another test. Almost overhead you will find the Great Bear, that faithful constellation which never deserts us. Look at Mizar, the second of the stars in the "tail". You may see a smaller star close beside it; the Arabs call it Alcor, and say that keen eyesight is needed to make it out – though your humble scribe can claim to see it without difficulty when clouds are absent.

◆   Note too Arcturus, with its lovely light orange hue; Ptolemy called it "golden red". Why did the ancients regard it as unlucky? Pliny even calls it "horridum sidus". Yet this is a shameful slur: look at Arcturus and admire its beauty – and do not forget the little semicircle which marks Corona Borealis, the Northern Crown.

Summer

We lose Venus and Saturn; we gain Jupiter. This summer, the brilliant planet, so aptly named after the king of the Roman gods, shines down from among the stars of Sagittarius, the Archer. True, from our climes it is low down, but even so it is truly magnificent. What kind of place is it? A world like our own Moon? It is further away from Mars, yet closer than Saturn, moving round the Earth in a period of almost twelve years. The Jovians, whoever they are, must be proud of their abode.

◆   Not far from the Archer is the heavenly Scorpion, with the red leader Antares – the "rival of Mars". In ancient lore the Scorpion caused the death of the hunter Orion by crawling out of a hole in the ground and stinging him in the heel. So when both were placed in the sky, Scorpius was put as far away from Orion

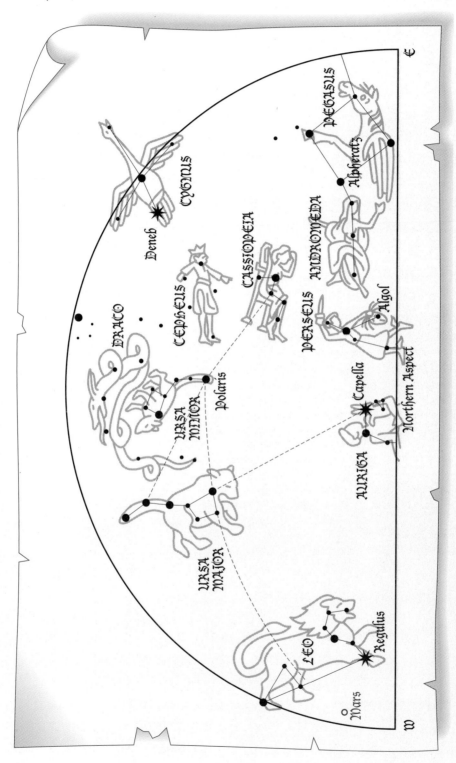

The stars of Summer, as seen from Winchester (latitude 50°) at 10 p.m. on 15 July 1000.

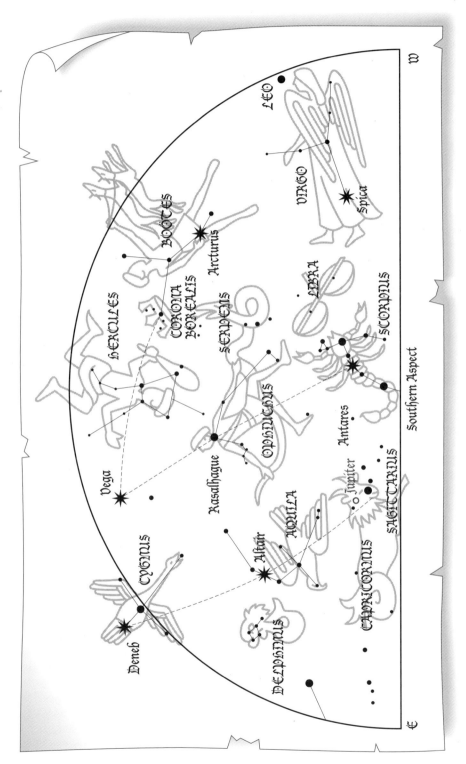

The stars of Summer, as seen from Winchester (latitude 50°) at 10 p.m. on 15 July 1000.

as possible, so that they can never meet again – and they are never above the horizon at the same time.

◆ Scorpius is magnificent, but do we see it in its entirety? It winds down to the far south. Travel to the southernmost part of our island, and you will just be able to glimpse the "sting", peering coyly up from the low-lying mists. Whether Scorpius extends still further south we do not know, but tradition has it that there are many bright stars which are forever concealed from us. In the Scorpion Al-Sufi notes a misty patch. Is this a cluster of stars, or something quite different?

◆ Three brilliant stars light up the evening sky in summer: Vega, Altair, Deneb. Of these, Vega is the leader, and it is steely blue, contrasting with the yellowish hue of Capella and the light orange of Arcturus. Vega lies in the constellation of Lyra, the Lyre or Harp, named for the instrument which Apollo gave to the great musician Orpheus. Aratus the Greek called Vega the Little Tortoise, because some lyres were made from tortoise-shells. Al-Sufi called it Al-Twazz, the Goose.

◆ Deneb lies in Cygnus, the Swan; the name comes from the Arabic Al-Dhanab al-Dajajah, "the Hen's Tail". But the stars of Cygnus make up an X, and the constellation is sometimes called the Northern Cross. Mariners who have sailed far across the sea have said that there is a brighter cross in the far south, but whether or not this is true I do not know. As for Altair, this leads the Eagle, a bird dispatched from Olympus to collect the shepherd-boy Ganymede, destined to become the cup-bearer of the gods. Why, I wonder, is it always thought to be an unlucky star, warning of danger from reptiles?

◆ Leo is sitting in the west; by August it and Mars will be gone. But Pegasus is rising in the east, and overhead we have the large constellation marking Hercules – though, truth to tell, Hercules is not very bright; one might claim that the great hero of mythology deserves better!

Autumn

As the leaves turn brown and the chill wind begins to blow, so the skies change. We see the Great Bear at its very lowest, though it is never lost below the horizon; almost overhead is the W of stars that is Cassiopeia, and between this and the Bear is the Pole Star, known either as Polaris or as Cynosura. This is the star making the northern pole of the sky, but it has not always done so. In ancient times this honour belonged Thuban, in the constellation of Draco, the Dragon; but Thuban is a much fainter star, so that navigators have profited by the change.

◆ Jupiter is now lost to the evening sky, and so are Mars and Venus; but Saturn is again with us, close to the reddish-orange Eye of the Bull, Aldebaran. We have too the Pleiades, and this means that Orion is not far behind. Before Winter has far advanced Orion will again be seen not long after sunset, and the Hunter will dominate the scene for the rest of the year.

◆ One strange fact should be noted. During August it often happens that we see flashes of light in the sky, known commonly as "shooting stars", though of course they cannot be stars. There have been past reports of great numbers of these objects during late October, and it has even been proposed that they are chips off the

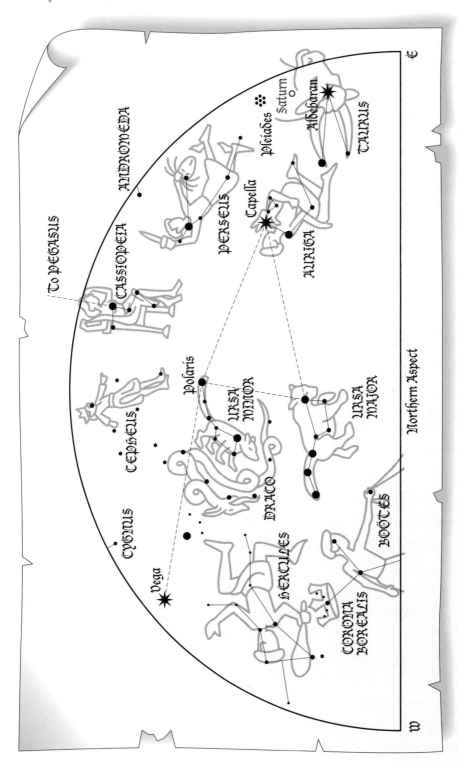

The stars of Autumn, as seen from Winchester (latitude 50°) at 10 p.m. on 15 October 1000.

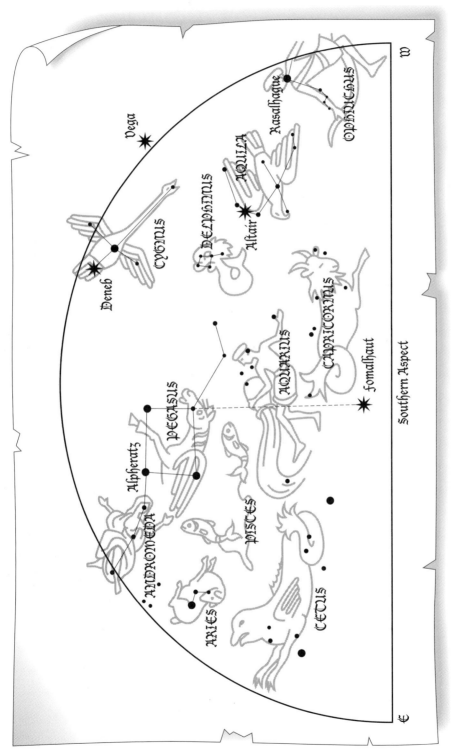

The stars of Autumn, as seen from Winchester (latitude 50°) at 10 p.m. on 15 October 1000.

firmament. Of course, they cannot come from afar. They must be due to vapours lifted away from the Earth by some agency about which we are ignorant; but they can be truly striking, and those who study the night sky may at any time be amazed by them.

◆ Lastly, do not omit to seek out Fomalhaut, in the little constellation of the Southern Fish. You will find it very low down, below Pegasus. The Arabs have called it Al-Difdi al-Awwal, the First Frog. It is white; some reports have claimed it to be red, but this is a mistake. It can sometimes appear orange or red because it is so low down, but in truth Fomalhaut has no colour at all.

◆ And so we come to the end of the year 1000 – the last year of the First Millennium. What great events are in store for us before this Millennium too ends, in the year 2000? Will men still inhabit the Earth, or will some disaster have overtaken us? Perhaps our fate lies in our own hands.

Marvels in the Heavens

◆ We say that the skies do not change. This must assuredly be true. Change indicates imperfection; surely imperfection cannot be allowed in the Heavens? Nevertheless, there are times when marvels are seen; we have yet to understand them, but in the name of true knowledge, why should we not chronicle them and submit them to interpretation?

◆ Records of the sky go back many, many centuries – long before the time of Christ our Saviour, long before the start of the first Millennium. There are numerous observations from China, from Mesopotamia, from Egypt and from Greece; some of these can be explained easily, while, we must shamefully admit, others cannot.

◆ Eclipses of the Sun and Moon not only occur, but do so with a certain rhythm, so that they can be predicted; did not an eclipse of the Sun, predicted by the first of the great Greeks, Thales, end a war between the Lydians and the Medes? And did not Nicias, commander of the Athenian armies in Sicily, delay evacuation because of an eclipse, so that he and his forces were destroyed and Athens itself brought to the brink of defeat?

◆ The Greeks were enlightened – even though there were a few, such as Aristarchus of Samos, who held

17

heretical views, and even went so far as to entertain the absurd belief that our Earth is a mere satellite of the Sun. What self-evident nonsense! Seated in front of this Yearbook, do you feel as if you are moving at a great speed through the Heavens?

◆ We may be sure that there will be no eclipses of the Sun seen from our islands during the year 1000, covered by this Yearbook, so that I will say little more about them here; but it would be churlish not to refer to some eclipses which are described in our great history, the Anglo-Saxon Chronicle.

◆ As our educated readers will be well aware, the Chronicle was due to the genius of King Alfred, our greatest monarch and no doubt the greatest monarch in world history. (In so saying I ask you to believe that I mean no disrespect to our present King, Ethelred II, who leads us in our never-ceasing battle against the barbarians from Denmark.) The Chronicle made its first appearance in the year 890, certainly under the supervision of King Alfred himself, and has been updated ever since.

◆ Issuing such a work is no easy matter, in these troubled times. It must be compiled and maintained by many scribes who (as I do) prefer to remain unnamed. Sometimes I dream that one day it will be possible to make many hundreds or even thousands of copies, so that all may enjoy them. But by what alchemy this feat might be accomplished I know not. Perhaps it will be achieved, by some means to be devised far in the future, as we learn more and more. For the moment, sadly, we must accept that each copy has to be compiled individually.

◆ Eclipses? Many are mentioned. Thus from 664, we read: "This year the Sun was eclipsed, on the eleventh of May; and Erkenbert, King of Kent, having died, Egbert his son succeeded to the kingdom."

◆ In 795; "This year was the Moon eclipsed, between cock-crowing and dawn, on the fifth day before the Calends of April; and Erdwulf succeeded to the Northumbrian kingdom on the second before the Ides of May."

◆ In 806; "This year was the Moon eclipsed, on the first of September; Erdwulf, King of the Northumbrians, was banished from his dominions; and Eanbert, Bishop of Hexham, departed this life."

◆ And in 827, "This year was the Moon eclipsed, on mid-Winter's mass-night; and King Egbert, in the course of the same year, conquered the Mercian kingdom."

Can it be that these eclipses mark great events – some disastrous? It may well be, but we are here discussing astronomy. Astrology is reserved for a separate Yearbook. We do at least know why these eclipses happen. Not so with other portents, as we learn from the accounts in the Chronicle. As your humble scribe, and no more that this, I present you with three strange and menacing cases:

◆ "774. This year the Northumbrians banished their king, Alred, from York at Easter-tide; and chose Ethelred, the son of Mull, for their lord, who reigned four Winters. This year also appeared in the heavens a red crucifix, after sunset; the Mercians and the men of Kent fought at Otford; and wonderful serpents were seen in the land of the South-Saxons."

Portents in the Anglo-Saxon Chronicle. A red crucifix (aurora?)
appears in the sky at sunset. King Elwald of Northumbria is
slain by Siga, after which Siga himself perishes. The Mercians
fought the Men of Kent at Otford; the South Saxons were
plagued by wonderful serpents; and fiery dragons were seen in
the skies.

◆ "789. This year Elwald, King of the
Northumbrians, was slain by Siga, on the eleventh day
before the Calends of October; and a heavenly light
was often seen on the spot where he was slain."

◆ "793. This year came terrible fore-warnings over
the land of the Northumbrians; terrifying the people
most woefully; these were immense sheets of light

rushing through the air, and whirlwinds, and fiery
dragons flying across the firmament. These   .
tremendous tokens were soon followed by a great
famine; and not long after, on the sixth day before the
Ides of January in the same year, the harrowing
inroads of heathen men made lamentable havoc in the
Church of God in Holy Island, by rapine and slaughter.
Siga died on the eighth day before the Calends of
March."

◆   Can we, in truth, question that these portents
were sent as warnings to us all? And what can be the
very nature of these "fiery dragons"? There is talk
that in the far north of our islands, in chilly Scotland,
these dragons - or "battles in the sky" as some think -
are often seen; they have been called the Northern
Lights, and it is said that when they are seen children
must be taken indoors for safety. Do the lights shine
in the upper part of the air? Do dragons fly so high
we can see only their glowing breath? We just do not
know.

◆   And in 800 we learn from the Chronicle that "a
cross was seen in the Moon, at Wednesday, at the
dawn; and afterward, during the same year, a
wonderful circle was displayed around the Sun."

◆   These wondrous phenomena were seen long ago; it
is from King Alfred, of glorious memory, that we know
about them.

Let us now turn our eyes toward other strange things
which appear to perplex us and alarm us. These are
what were called by the Chinese "hairy stars" - in our
tongue we refer to them as comets or comet-stars. Yet
they are not stars; they appear, shine balefully down

upon us, and vanish into the blackness as softly and as quickly as they appeared. That they are portents of major disaster can hardly be questioned, but of their true nature we remain ignorant. What can a mere scribe such as myself say about them? First, let us look back at the Chronicle.

◆ "678. This year appeared the comet-star in August, and shone every morning, during three months, like a sunbeam. Bishop Wilfred being driven from his bishopric by King Everth, two bishops were consecrated in his stead."

◆ "729. This year appeared the comet-star, and St. Egbert died in Iona. This year also died the atheling Oswald, and Osric was slain, who was eleven winters King of Northumberland."

◆ "995. This year appeared the comet-star." And who, pray who, can erase the memory of those fearsome raids from across the water, made during that year? Can it be that these comets are the servants of those fierce barbarians from Denmark, and are sent to cast gloom and terror across our own fair land? Surely it is not for me to pronounce.

◆ We do not base our knowledge wholly upon the Chronicle. There are stories about a great comet-star seen in the year 837, whose fiery tail stretched wholly across the firmament, and which became so brilliant that it cast shadows, turning night into day. There is, too, the writing of the monk Ralph Glaber, still happily with us on earth, who entered a Burgundian monastery in 997, and has provided us with all we know about the great comet-star of a year of fearful Danish attacks which the forces of King Ethelred were unable to repel.

Hairy stars, or comets, strike fear into the Anglo-Saxons.

◆ **I must repeat here what Glaber has told us:**

It appeared in the month of September, not long after nightfall, and remained visible for nearly three months. It shone so brightly that its light seemed to fill the greater part of the sky, then vanished at cock's crow. But whether it is a new star which God

launches into space, or whether he merely increases
the brightness of another star, only he can decide ...
What appears established with the greatest degree
of certainty is that this phenomenon in the sky
never appears to men without being the sure sign
of some mysterious and terrible event. And sure
enough, a fire soon consumed the church of Saint
Michael the Archangel, built on a cape in the ocean
which had always been the object of special
veneration throughout the whole world.

◆ What, then, are comet-stars? If sent from afar,
beyond our air, then this would mean that the skies
would change; we know from the teachings of the
Church that this cannot happen. There is only one
explanation: they must be due to fiery vapours sent up
from the Earth itself. We cannot tell when they will
appear and terrify us, and we must wait, with
apprehension, to see whether a new portent will
become visible during the year 1000, to usher in the
new Millennium. We can only offer up our prayers
that this will not happen.

◆ With this I close my modest account of marvels in
the heavens. I pray that before the end of our new
Millennium, in the year 2000 – how far ahead that
seems! – we may learn more about the meaning of
comet-stars, and perhaps even allay the fears which
now surround them.

The Astronomical Achievements of the First Millennium

by Albert Pendleburiensis, Archdeacon of Lancastria

◆ I send my greetings with this Yearbook for our second Millennium to my old master and encourager Gerbert of Aurillac, who in this 999th year since the birth of Christ Jesus has become his Vicar on Earth, or Pope, and has taken as his title Pope Sylvester II.

◆ For when His Holiness was Master Gerbert, some 30 years ago, he travelled to Spain as a scholar, astronomer and man of great learning, to discuss many matters with the Mohammedan learned men who live in that country. I, as a young monk from Lancastria in the land of Anglia, then a subject of King Edgar who had been given leave to study in the school at Rheims in France, was accorded the honour of being allowed to accompany Master Gerbert to Spain. I later assisted Gerbert in composing his books, especially those dealing with astronomy and dial-

A.C.

Gerbert of Aurillac, Master of the Schools at Rhiems,
Archbishop of Ravenna and, in AD 999, Pope Sylvester II. This
Pope, who saw in the Second Millennium, encouraged the
mathematical sciences along with the use of instruments such as
the abacus, globe and astrolabe; numbers, after all, lay at the
heart of God's Creation.

making. Accordingly, I am well suited to write upon
our current state of astronomical knowledge in 999,
having travelled across Europe twice, visited the
astronomers of Mohammedan Spain, and journeyed to
Byzantium and the Golden City of Constantinople,
where they speak the Greek language, and have full
and complete astronomical books in Greek of which we
in the Latin West have only fragments in translation.
It was in Constantinople, indeed, that I myself learned
Greek, and was thereby able to study books unknown

to other men in Anglia, Francia, Germania, Hibernia or Scotia.

◆ Though we in Northern Europe are not so advanced in astronomy as the peoples of Spain, Byzantium, and the Mohammedan cities of Damascus and Cairo, and we have fewer books, and are much troubled by the barbarous Danes, there is clearly a growing revival of interest in astronomy here. This revival began some 200 years ago, and I am proud to say that its pioneer was a native of our own island of Anglia.

◆ This was the Venerable Bede, a monk of the once great Abbey of Jarrow in Northumbria (the Danes, alas, sacked it in 798), and I am indebted to the most learned and revered Sister Beata of Nunnaminster for an account of his life and works. Brother Bede, who died in AD 735 at the fair age of 62, never left Northumbria, though he was sent many books for the Abbey library by his friend Abbot Benedict Biscop. Abbot Benedict travelled to Mediterranean lands on Abbey business, and via this source Bede became acquainted with the learning of the South. This enabled Bede to become the first truly Anglian astronomer, able to make advances on what was known in Italy. He wrote upon time, calculation, and on several aspects of the Heavens and the Earth. It is even said that he made a stellar globe, so that he could teach practical astronomy to others. Perhaps most significant of all, Bede devised improved formulæ, now used across much of Latin Christendom, whereby we can calculate the correct date of Easter Day. This most important of all our Christian festivals is calculated each year from the motions of the Moon around the time of the Spring Equinox, and is a task of great complexity, demanding a detailed knowledge of practical astronomy and calculation.

◆   Born in the same year as that in which Bede died, 735, was another great Anglian scholar, Alcuin of Eboracum.[1] Alcuin was the greatest teacher of his age, first teaching astronomy, calculation, Latin grammar and other subjects in his native Abbey in York, before going to live in Continental Europe. Indeed, when Charlemagne finally succeeded in bringing peace to Europe and becoming Emperor of the West, he set up several monastic schools and places of learning. He invited Alcuin to leave York and take charge of his education system as "Master of the Palace Schools", and teach within the Imperial Court. Alcuin became Charlemagne's chief scholar, and an authority on every branch of learning and science. During this first Millennium, therefore, astronomy in Anglia and Europe, and in the lands of the Mohammedans, has been brought very close to achieving perfection.

◆   For centuries beforehand, however, the pagan people of Egypt and Babylonia had feared and worshipped the Sun, Moon and stars as gods and goddesses, though, as the ancient Jews had realised by the time of King David in 1000 BC, these were only inanimate objects, moving under the guiding hand of God their Creator. For in his Psalms, King David speaks of the "Lights of the Firmament" as natural as opposed to supernatural bodies. And by 600 BC Thales the Greek had come to appreciate that the Sun is but a

---

[1] The Roman name for York.

"fiery stone" and that the Moon is merely a ball which shines by reflected light.

◆　It was, indeed, the Greeks who first placed astronomy on a proper footing. For by noting the heights and angles of the Pole Star in different places, the motions of the planets amongst the stars, and the great cycles that produced eclipses, they showed that astronomy was not a subject of fearful superstition, but one in which man could exercise the greatest natural aptitude which God had given him: his reason. For as we now know and teach, astronomy is one of four sciences of proportion – called in the Latin tongue the Quadrivium – which show us how, by the careful use of reason, we can explain the logical interconnectedness of so many natural phenomena; the other three sciences are geometry, arithmetic and music. I shall say more of these sciences anon.

◆　It is unfortunate, alas, that the writings of so many learned Greeks are not available in their complete editions to the scholars of Anglia, and indeed of Latinate Europe. The mischief wrought by Alaric, the Visigoths and the other barbarian hordes which overran much of Europe some 500 years ago destroyed many fine libraries in Italy, Southern France and the cities of the Adriatic Sea. And likewise, when the Mohammedan religion came into being 378 years ago, in the year AD 622, the Caliph Omar destroyed the great library in the then Christian city of Alexandria in Egypt, though soon after the Mohammedans became great students and great preservers of astronomy.

◆　As a result of these losses, what we have in our own libraries, such as those found in the monasteries of Rheims and Cluny in France, St Gall in Switzerland, Monte Cassino in Italy, and Winchester and Canterbury in Anglia, are extracts, fragments and digests preserving the ideas of Aristotle, Hipparchus, Aratus and other ancient astronomical writers, and most of all, of Claudius Ptolemy. We must therefore thank such men as Cassiodorus, Boethius, Macrobius and Isidore of Seville for collecting so many useful fragments and translating them into the Latin tongue, producing encyclopædias which bring together much useful learning in science and philosophy.

◆ The man who most clearly brought astronomy to perfection, however, was Claudius Ptolemy of Alexandria, who worked some 160 years after Christ. Indeed, Ptolemy himself was a great collector and encyclopædist of astronomy, who preserved, in his turn, the names and achievements of several earlier Greeks who are now known to us only through him. Ptolemy wrote several great books, none of which, alas, we possess in our libraries here in Northern Europe, but of which there are copies in the libraries of Byzantium and Andalusia, in the south of Spain. The first of these was his Magna syntaxis, or "Great Astronomical Compilation", which the Mohammedans call Almagest, which, so I understand, means the same thing in Arabic. This great book dealt with all aspects of the celestial motions. Ptolemy's other book was his Geographia, which treated of the science of making true maps of the Earth from astronomical observations. While I have read Greek copies of these books in Constantinople, and understand that the Mohammedans of Damascus and Cairo have them in Arabic translations (indeed, I saw an Arabic Almagest, with all its complex drawings, when I was in Grenada, though I could not read it), we Christians of the West possess only Latin fragments.

◆   Ptolemy's achievement lay in explaining the whole of astronomy in accordance with a clear set of principles. By setting up a pair of large brass rings, one in the plane of the ecliptic and the other on the true meridian, and equipping each with 360° divisions, he produced a form of armillary instrument. The ecliptic ring enabled him to track each planet across the sky with relation to the stars, and thereby keep an accurate record of their wanderings. The meridian ring gave Ptolemy his noon point, allowing him to find his time and relate the planets and stars to a fixed point of observation. Ptolemy devised other instruments which will be described elsewhere in this yearbook by my friend, Father Allanus Salfordiensis.

◆   By observing the heavens with his instruments, and comparing the places of the Sun, Moon and planets in his own day with those in the days of his forebears, most notably Hipparchus, Ptolemy was able to perfect astronomy as a predictive science. From the observations of ships' masts, lunar eclipses, and the Pole Star, and from shadows, Ptolemy knew that the Earth was a sphere, just like the Sun and Moon. (There are, however, some folk in our own day, no doubt guided by the Etymologies of Isidore of Seville, who still maintain that the Earth is flat.) Ptolemy also understood that, as all planets move at fixed speeds, one can arrange them in order of distance from the Earth. As the Moon rotates around us in 28 days, she must be much closer than dull Saturn, whose orbit around us takes up $29\frac{1}{2}$ years. Thus we know the order

of the planets: the Moon (28 days), Mercury (88 days),
Venus (267 days), Earth (365¼ days), Mars (2 years),
Jupiter (12 years), Saturn (29½ years). The heavens,
therefore, move at a perfect speed and in a strict
proportion that has not changed since Adam and Eve
were driven out of the Garden of Eden.

◆　　Philosophers have often been struck by the
contrast that exists between the perfection of heavenly
motions and the confusion we find on Earth. For the
heavens never actually change, moving as they do in
long and short cycles, whereas nothing on Earth is
ever predictable. Indeed, the random plagues, storms,
lightnings, earthquakes, famines, floods and droughts
of this world derive from the fall of Mankind, when,
through curiosity, Eve was tempted to eat the
forbidden fruit of the Tree of Knowledge in the Garden
of Eden. All earthly mischief stemmed from this act,
but while Eve's action brought chaos into the world, it
never affected the heavenly bodies. This, we know, is
why the Heavens retain their perfection.

◆　　And yet, you might argue, the Heavens do change;
for what of the lunar phases, comets, shooting stars,
and what the ignorant call "battles in the sky"? Well,
Eve's sin of disobedience tainted not only the globe of
the Earth – setting the four Elements of Earth, Air,
Water and Fire into a state of permanent warfare with
one another – but also the whole of the region below
the Moon: the sub-lunary sphere. This is why the
Moon shows phases, or changes. On the other hand,
Eve's damage to the Moon was peripheral, for the
Moon's changes are themselves perfectly predictable.
Yet, just like storms, plagues and thunders, other types
of aerial change cannot be predicted. Who can tell

when, as Aristotle argues, the stinking airs that rise
up from a battlefield or a swamp will catch fire from
the heat of the Sun, to produce a fiery star, or comet,
in the sky? For are not comets meteorological bodies in
the air, above the Earth, yet lower than the spheres
of the planets? Likewise, who can know when a great
wind will not sweep up stones into the sub-lunary
region, only to drop them as what the foolish call
"shooting stars"? The motions of the planets
themselves, however, never change.

◆　A major contribution of Ptolemy's to astronomy
was to explain why the planets move as they do. He
envisaged the Universe arranged as a set of nine
concentric spheres. (Ptolemy, of course, did not invent
this system, for it is said to go back to Eudoxus, and
other of the ancient Greeks, whose writings we know
only by report, but which Ptolemy incorporated into
his own works.) In the centre is the Earth, and around
us rotate nine spheres which, like the skins of an onion,
nestle exactly inside one another. Being perfect, each
sphere is made of a pure crystalline substance. The
first seven of the spheres are indeed transparent, and
each carries a planet – the Moon, Mercury, Venus,
Sun, Mars, Jupiter and Saturn – while the eighth is
black, and carries the fixed stars around us, once a day.
Beyond it is the invisible ninth sphere, or Primum
mobile (prime mover), which governs the speeds of all
the rest.

◆　Now for centuries, astronomers had never been
able to explain the complex movements of certain
planets: why, for instance, did the Moon wander
around the sky so much over her 19-year cycle, and,
most puzzling of all, why did the planets Mars, Jupiter

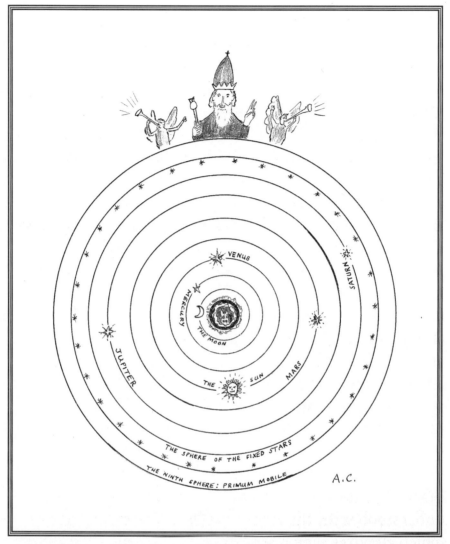

The mediæval Universe. At the centre of Creation is the Earth, composed of earth, water, fire and air. At the Earth's centre is hell, complete with Satan. Around the Earth revolve the Moon, Mercury, Venus, the Sun, Mars, Jupiter and Saturn, the fixed stars, and the Primum mobile, or prime mover. The whole is presided over by God, as king and Bishop of Creation. His angels praise Him everlastingly.

and Saturn display "retrograde" motions, wherein they seem to reverse backwards along their orbits? Ptolemy proposed that these motions could be explained if the shining planets were situated upon the rims of invisible circles. The centres of these circles – around which the planets rotate at a perfect speed – themselves rotate upon the crystalline spheres. Yet these two perfect motions, when viewed from the Earth, create the impression that the planet has suddenly started going backwards; thus he explained in theory what one could see in fact. By this device, Ptolemy made it possible to marry observations made with instruments to a doctrine of cosmological perfection, yielding a predictive practical astronomy which enables men to know where a given planet will be amongst the constellations for years to come. This is why, in our First Millennium, astronomy has achieved such a perfection, removed as it is from ancient superstitions and heathen deities.

◆　　I mentioned above that, in this First Millennium, astronomy, alongside three other sciences, has come to make up the Quadrivium of the Liberal Arts and Sciences. Each of the four parts of the Quadrivium, indeed, explains an aspect of perfection that can be understood with the God-given gift of reason: the gift which elevates men above all the beasts. For the first science, arithmetic, teaches us the logical power of numbers. Is it not wonderful that 2 plus 2 always equals 4, and that when an even number is divided by 2, the two parts are always identical? In fact, it was the learned Plato, some four centuries before Christ, who first showed us that number was the key to all rational knowledge. Alas, none of Plato's complete books survive in Anglia, though they do in Greek Byzantium and probably in Arabia. None the less, our

fellow-countryman, the Venerable Bede, wrote on calculation, and some of his rules are still in use today.
◆ Plato, likewise, spoke of the perfection of geometry, which is the next Liberal Science. Although the actual works of the Greek Euclid are not, sadly, known in the Latin West, we are fortunate to have compilations on geometry by Boethius and others. Is it not amazing to learn that the internal angles of a triangle always add up to 180°, and that a square on the hypotenuse of a right-angled triangle is always the

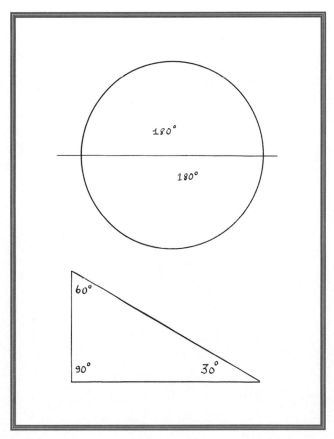

The wonders of geometry: a circle divided through its centre always produces two 180° semi-circles, while the internal angles of a right-angled triangle always add up to 180°.

same in its area as the squares on the two other sides? For here we find a deep philosophical truth that approaches the forms of perfection discussed 1,400 years ago by Plato.

◆ Our next Liberal Science is music, which deals not so much with the art of playing upon an instrument as with the geometry of beautiful sounds and mathematical sequences. It was indeed Helena, the

Helena, the Cantrix of Paris, with the instruments of musical proportion: bells, monochord, harp and organ.

wise and most learned Cantrix of Paris, who first
made me aware of the perfection of music, many years
ago, when I was a student of Gerbert. If a string upon
a harp is tuned to play a particular note, and then we
stop, or put pressure on, that string with a finger
exactly in its middle, then the half-section of the string
will sound precisely one octave above the full length.
The same applies to flutes, pipes and organs. In this
way, the sequential vibrations in the air combine into
chords which delight or offend the ear. Those chords
which delight are made up of strings or pipes that are
arranged in an exact mathematical sequence of length,
whereas discordant sounds derive from irregular
mathematical sequences. In this way, that perfection
to which all men aspire makes us respond joyfully to
the sound produced by perfect mathematics, whereas
we grind our teeth at an asymmetrical sound. And, as
ancient Orpheus well knew, music can charm not only
wild beasts, but even the souls of the dead in Hell;
while the young David was able to soothe the madness
of King Saul by his beautiful playing and singing.

◆   And so the Science of astronomy treats of
perfection in the Heavens, as the motions, phases,
conjunctions and ever-changing angles of the stars and
planets possess changeless wonderful sequences in
their turn.

◆   In this way, the four sciences of the Quadrivium
combine to teach us how the perfection of mathematics
and the abstract beauty of numbers, the forms of
geometrical figures, and the sounds that touch the soul
of man are all brought together in the very perfection
of the Heavens. Thus are we brought to the noblest
goal of astronomy, cosmology, which explains the

unity of everything, encompassing as it does the world, the planets, the mind of man, and the celestial majesty of God.

◆ It is true that over the ages astronomers and philosophers have drawn up differing models for the arrangement of the world. Ptolemy saw the globe of the world as divided into four quarters, the whole being set at the centre of a sequence of rotating crystalline spheres. Conversely, Bishop Isidore of Seville, who lived some 400 years ago, saw the Earth as a flat wheel set within the planetary spheres. Everyone agrees, however, that beyond the Earth the physical universe is very vast, and that the Earth and mankind occupy but a tiny fraction of its space, as Boethius tells us: "Thou hast learnt from astronomical proofs that the whole Earth compared with the Universe is not greater than a point, that is, compared with the sphere of the Heavens", and that of the Earth "only one-quarter, according to Ptolemy, is habitable to living things".

◆ We can tell that the Moon is many thousands of miles away from us by observing the slow passage of the Earth's shadow across its face during a lunar eclipse. And then, during a total eclipse of the Sun, we find that the Moon exactly blots out the face of the Sun. It is reasonable to assume, therefore, that the Sun's sphere or orbit around the Earth must lie at a distance that is greatly beyond that of the Moon, encompassing as it does the orbits of both Mercury and Venus. Consequently, the Sun must be the largest, as well as the brightest and hottest, of the planets.

◆ Although the Sun and Moon display visible disks in the sky – the Moon's like a blotched silver plate, and the Sun's perfectly golden – they are none the less the only astronomical bodies to present disks to our eyes, for Mercury, Venus, Mars, Jupiter and Saturn, just like the stars, are but points of light.

◆ And yet I wonder. When I was travelling in Spain 30 years ago, I met a young Mohammedan who, in the dry climate of Andalusia, assured me that he could see points, or "horns", upon Venus. He was a medical student from the great hospital in Grenada, and he was particularly interested in the delusions of the mad. Whether he had especially acute vision, which revealed to him things him which other men could not see, or whether he had been touched by the delusions of his patients, I was never able to establish. On balance, I choose to believe the young man to be gifted with wonderfully sharp eyes.

◆ Yet true cosmology deals with more than the mere arrangement of objects in space. It deals, indeed, with the beginnings and endings of things.

◆ Plato in his dialogue Timaeus (parts of which do survive in Latin translation in our Anglian libraries) spoke of the Divine Creator making the Universe in a similar way to that in which a potter makes a pot – to a design. Plato's ideas, pagan as they are, fit nicely with our Christian ideas of Creation as recounted in the Book of Genesis. Plato's pupil Aristotle, however, said that the Cosmos was probably timeless, having no obvious beginning or end, but consisting only of eternal cycles. This view, I have found on my travels, is repugnant not only to Christians, but also to Jews and Mohammedans, all of whom see an act of Divine Creation as lying at the start of the Cosmos.

◆ Cosmology also deals with the very nature of time itself, for scholars have long since wondered whether the time which we experience on Earth - measured by the rising and setting of the Sun - is the same as that experienced by the unborn, the dead, the Devil in Hell, the Angels in Heaven, and God Himself. For neither Heaven nor Hell lies within the realms of sunshine, existing as they do in different dimensions of being. Hell lies within the very bowels of the Earth, and is fuelled by fires which never cool, to generate torments that never cease. Heaven is beyond the spheres of the visible Universe, in the infinite and wonderful realm of light that lies beyond the starry sphere, where God reigns with His angels, the saints and the souls of the blessed dead. This is indeed why we speak of the souls of the wicked as descending into Hell beneath the ground, and those of the virtuous as ascending through the sky to Heaven.

◆　In neither Heaven nor Hell, therefore, can normal time exist. For in Hell, the fiery torments will last for ever, so that a second feels like a year, while those who enjoy God's blessings in Heaven can never change, for perfection can never alter. In an awareness of time there is, after all, a component of expectation, as we wait for things to change. Yet, as being in the presence of God can provide no imaginable improvement, Heaven can have no future expectation, and hence, no sense of time! Consequently, time is not an absolute thing, but is relative to where you are in the Universe. Many scholars have written about time, but it was the learned St Augustine, working some 600 years ago, who produced one of our foremost discussions on time, and its relationship to cosmological infinity, in Book XI of his Confessions. I would recommend this book to all who are interested in the nature of relative time and space, as it is easily available in most of our major libraries.

◆   The nature of time in its cosmological setting has always fascinated me, and in my travels between Spain and Byzantium, Rome, Rheims and Durham, I have talked to astronomers who were Christians, Jews and Mohammedans upon this matter. I have also read many books in Latin and Greek, and some that were translated from Arabic. And the more that I read and think, the more I wonder how the astronomers of the New Millennium will perfect their ideas. How, indeed, in 1,000 years from now, shall we talk of cosmology, and of time, within the infinite vastness of Creation, and with what instruments shall we view it?

◆   By then, I shall have long departed to taste for myself the pleasures - and hopefully, not the pains - of infinity. And for the short time that I now have left upon this Earth - for I am 50 years old, my teeth are worn down, my joints are sore from too much walking and my eyes are no longer clear when I read - I truly hope that my old Master, the new Pope Sylvester, might translate me from my Archdeaconry of Lancastria, where I am obliged to walk through so many bogs, in the rain, to visit the isolated churches of my native county, to a well-endowed bishopric, so that I may prepare for my departure from this world in comfort.

Astronomical Instruments of the First Millennium

by Allanus Salfordiensis, Prior of the Monastic House of St. Frideswide at Oxenforde

◆ Ever since astronomy ceased to be confined to superstitions about the celestial deities of the pagans, and emerged as a Liberal Science, it has used instruments to measure the angles and motions of the heavenly bodies. One of the most important of all classes of instruments is that of those used to find time.

Sundials

It is said that Berosius, some 300 years before the time of Christ, made a sundial in Babylonia. This dial was supposedly carved as a quarter of a circle into a block of stone, so that, when it was set up facing south, the shadow of a bead or a pointer, called a gnomon, designated the hour upon its engraved lines. So successful was this design that it was used by many peoples around the Mediterranean, where it came to be called scapha by the Latins from its resemblance to something "scooped out". Such scapha dials are also used in the Mohammedan world at the present day.

◆ For a variety of reasons sundials are very awkward things to construct, which makes the accurate reckoning of time very difficult. The first of these difficulties derives from the changing angle of the noon-tide Sun in the sky at different seasons of the year. Unless one has a scaphae dial that is set in the plane of the world's equator, it is very hard to compensate for these changing angles. The second difficulty arises because of the varying amount of daylight striking the Earth between December and June. For though we divide our days into 24 hours, we always reckon by 12 hours of daylight and 12 hours of darkness. At mid-Summer, therefore, we have 12 long daylight hours and 12 short night hours, which are reversed in Winter, so that only at the Spring and Autumn Equinoxes, in March and September, do we have 12 equal hours of day and night. These unequal hours inevitably make sundials difficult to devise, and constitute a great mischief, especially to those who try to construct ways of marking time by means of inventions.

◆ One modern scholar who has addressed himself to the construction of accurate dials, the correct determination of the angle of the Pole Star to fix the gnomon, and the angular displacement of the hour lines, is Master Gerbert of Aurillac, our new Pope Sylvester II. I had the honour of studying under Gerbert at the Cathedral School at Rheims, in France, and have been assiduous in bringing as much of his astronomical learning as I can (some of which he gained from the Mohammedans in Spain) into our island of Anglia.

◆ Sundials, of course, need not be of the simple scapha form, but can also be set up either horizontally or vertically, though the calculation of the lines is much more difficult when constructing dials with flat surfaces. Even so, I have seen several flat-plane dials in our island of Anglia, and have constructed several myself. Indeed, it is my fond hope that this learning might well flourish and grow in this City of Oxenforde, which is a well-defended and prosperous town, with many fine monasteries full of ingenious young monks and priests.

Horologia, or Clocks

Our blessed and much-loved King Alfred, over 100 years ago, devised an ingenious way of measuring the time when the Sun and stars were not visible. Knowing the rate at which a candle burned, he had equal divisions and numbers cut into one, and found that it made an excellent night-clock. Alfred's invention is now in widespread use.

◆ Horologia, or artificial devices for telling the time, really began about 800 years ago, when Ctesibius of Alexandria had the idea of using a regular flow of water to measure time. His device was called – in his native Greek – a clepsydra. By the dripping of water at a regular rate from one vessel to another, through a very small hole, a float is made to rise. By attaching a weight and a piece of string across a small pulley, or else using a simple toothed wheel, it became possible to make the rising water turn a pointer, which was then made to read off the hour of the day or night from a circular dial with 12 divisions.

◆ Indeed, in our cloudy climate it is often necessary to have some way of marking the hour, especially at

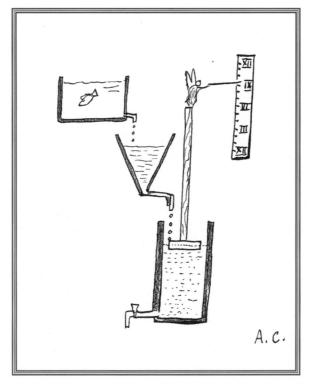

A simple clepsydra, or water clock. Water from a pair of small cisterns (the funnel-shaped one helps to equalise the water pressure) drips slowly into a cylinder. This gradually raises a wooden float, on top of which stands an angel holding a pointer, marking off the hours on a scale. Every 12 hours, the cylinder has to be emptied, and the cisterns refilled, in order that time-telling can begin afresh.

night, for otherwise the daily and nightly services of our monasteries could not be observed at the correct times. In my own monastery of St Frideswide, we have a lay-brother, Simon Stultus,[2] a slow-witted, gluttonous fellow, whose sole task is to ring the prayer-bell at the correct hours, depending on the position of the Sun or stars in relation to particular

---

[2] Literally, "Simple Simon".

roofs or towers. Yet Simon often absents himself
without leave to go guzzling in the shop of Lilia the
pie-maker, and then falls asleep. I therefore set about
devising a clepsydra with a large wheel, upon the outer
rim of which would be placed 12 heavy brass balls,

Simon Stultus dozes with a half-eaten pie. Behind him,
Abbott Allanus's water clock is in danger of ceasing to operate,
unless Simon replaces the brass balls in their swivel-cups,
and tops up the cistern that feeds the water escapement.

each about the size of a hen's egg. As the wheel rotated through each 12th of a revolution, a ball would fall down and hit a metal gong, to loudly mark the hour. If this advanced clepsydra works as I hope it will, Simon can be employed in more appropriate duties, such as assisting the privy cleaner, Lay-Brother Daniel Cloacinus,[3] though Simon will still have to gather up the fallen brass balls every few hours and replace them in their indentations on the wheel.

◆ Many people, in fact, have applied their minds to trying to tell the time by means of inventions or machines. Pacificus, Archdeacon of Verona in Italy about 150 years ago, is credited with several such devices. Some of them were clepsydras, though I have been told in a letter from my friend Aelfred Cliftoniensis, who has spent some time in Italy on the Archbishop of York's business, that others tried to harness the force of falling weights. Although I have heard of several men who have tried to measure time by means of falling weights connected to trains of toothed wheels, these can never be made to work properly until some ingenious fellow devises a way of enabling the weight to fall slowly rather than crashing down the tower! One experiment with such a weight-clock was made in a monastery not far from here some years ago, but it led to its deviser being remembered not as he had been known in life – as Brother Aethelwulf Ingeniosissimus,[4] but as Aethelwulf Contusissimus.[5]

---

[3] Literally, "Dan the Lavatory Man".

[4] Literally, "Aethelwulf the Supremely Ingenious".

[5] Literally, "Aethelwulf the Completely Squashed".

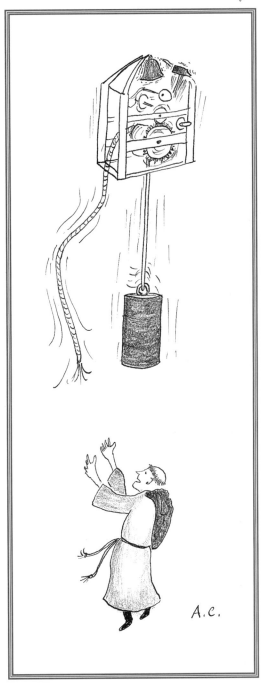

Brother Aethelwulf Ingeniossimus about to become Aethelwulf Contusissimus, having failed to supply his clock with an effective escapement to control the fall of the driving weight.

A.C.

◆ Yet one of the major problems faced by all those persons who wish to mark the hour by inventions is our universal use of hours of unequal length at different seasons of the year. For all the inventions that I have seen or heard of, be they driven by water or by weights, run at a constant speed and must be constantly adjusted to work at different seasons. Indeed, if measuring the time by means of inventions ever becomes widespread, we shall all be obliged to regulate our lives and prayers by 24 equal hours, with 18 hours of daylight and 6 of darkness in June, and the reverse in December.

Observing and Teaching Instruments

It was Claudius Ptolemy, about 840 years ago, in his *Magna syntaxis*,[6] who described a set of instruments used to measure the risings and settings of the stars. Though no copies of this book are known in our Latin West, I was privileged, when a pupil of Gerbert at the School of Rheims - of which he was then Master - to be shown a long letter. It had been sent to him, I think, by one Gregory the Archimandrite, a senior Christian priest of Constantinople, who, unlike most Byzantines, wrote moderately good Latin. I was permitted to copy out parts of the letter, and it is from this record which I have before me that I write about the instruments in Ptolemy's *Magna syntaxis*.

◆ Ptolemy describes his great brass rings, called armillaries. The main ring was adjusted to the plane of the ecliptic, or zodiac, and against it one could measure the Sun's course in the sky and relate star positions to the same coordinates. In summer the Sun moved above this ring, and in winter below it, as the position of the

---

[6] The Almagest.

Sun on the ecliptic changes with the seasons. A vertical ring could also be added to designate the meridian, so that when both rings were used in conjunction, it was possible to denote the position of any celestial body with regard to another. This ecliptic armillary has since been developed so that a third ring, that of the celestial equator, can also be added, along with a celestial pole, so that all manner of motions and coordinates can be shown.

◆   Ptolemy also knew the use of the quadrant, or quarter-circle, divided into 90°, by means of which one could measure the height of any star or planet in the vertical, or azimuth.

◆   He also described the principles of the astrolabium or astrolabe, a most ingenious jewel of an instrument capable of solving a multitude of mathematical problems. I understand that Gerbert obtained one out of Spain, but I never saw it. But I shall say no more about the astrolabe, partly because Gregory the Archimandrite spoke of only it in passing, and because I understand that Brother Aelfred Cliftoniensis has sent a detailed description of such an instrument from Italy, and that it will be appearing in this Yearbook.

◆   Yet one ingenious instrument which Gregory described was the "parallactic rulers" of Ptolemy. They were devised, I understand, for measuring the height of the Moon at critical stages of her orbit, though, in fact, they can be used to measure any astronomical altitude.

◆   The rulers work on the principle of the "rule of three", namely, if three rods of the same length are attached by their ends, they will form an equilateral triangle with a 60° angle in each corner. If one of these

An observer using Ptolemy's "parallactic rulers" to measure the altitude of the Pole Star as seen from York.

rods is set exactly into the vertical with a mason's plumb-line, then one can use the other two rods to denote an exact 60° angle in the sky. Yet if only two of the rods are joined, by means of brass swivel pegs which enable them to move so that the end of one rod passes across the length of the other, then it is possible to employ the rods to denote any angle less than 60°.

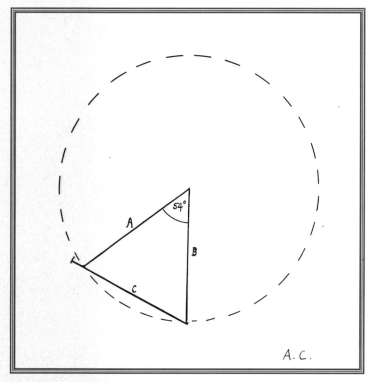

A. C.

The principle of Ptolemy's rulers. The three rods, A, B and C, are all of the same length, which is equal to the radius of the circle. If all three were joined at their ends they would form an equilateral triangle, with angles of 60°. But if A and B, as radii of the circle, are bridged by C as shown, with the end of the rod A joining C some distance from its end, then C will form a chord of the circle which subtends an angle of less than 60°. From a pre-calculated table of chords to a circle, the length of the chord on C, as marked by the end of the rod A, can be used to measure the resulting angle as a fraction of a complete circle. In the present example, the angle is 54°, corresponding to the latitude of York.

◆   Now, if one divides each rod into 100 equal parts, one can create, from the crossing point upon the rods, a set of proportions, let us say 100 : 100 : 50, or 60° from the horizon. By simple calculation, therefore, one can establish the angle denoted at A on the drawing from these proportions. One can, moreover, measure an angle in this way without need for the awkward procedures that are necessary to divide up a circle, for it is much simpler geometrically to divide up a straight rod into a given number of equal parts than it is to divide a closed curve, such as a quadrant.

◆ When I teach astronomy to our young monks at St Frideswide's, I also make use of a celestial globe, which I constructed from a ball of lathe-turned wood, a foot in diameter, covered with gores of fine parchment, and painted blue. I next drew the equator, ecliptic, and meridian lines upon it in gold, and the stars in silver after the correct manner of the 48 constellations of Ptolemy, as far south as the Tropic of Capricorn. I was guided in its design by a globe which Gerbert himself had used to teach me astronomy at Rheims, although the celestial globe is by no means novel to our own time. The Venerable Bede was said to have had one at Jarrow 200 years ago, while even the Romans were believed to have been acquainted with its use.

◆ I also devised a "planetarium" in wood and metal so that I could teach my pupils the way in which the Heavens move around the Earth. This instrument, however, was not of my own invention, but modelled upon that with which I had been taught over 20 years ago at Rheims. Likewise, I also learned of the abacus from Gerbert, and I in turn not only employ it for my own calculations, but also teach its usage to my own pupils. When handling large numbers, such as the multiplication of CCLXVIII by LXXIV, and especially when calculating the date of Easter from the Spring or Paschal Moon, it is much faster and more accurate to slide the small wooden balls upon my abacus than it is to work it out laboriously on a writing slate.

On the Nature of Light

Many learned men have been fascinated by the nature of light, which is the only natural agent that comes down from the Heavens to the Earth. Because light has its origin with God in the Heavens, it is by nature pure and white. When, however, it enters our sub-lunary world, which is beneath the Moon, it produces that phenomenon which I understand Aristotle calls meteorologia, by which it is inevitably corrupted by the sin of the world.

◆   As a consequence, the white light decays, and produces a spectrum of colours. We most clearly see this corruption or decomposition if we observe light as it passes through a drop of water or a piece of crystal, to produce a blur of reds, yellows and blues. But I shall write no more about light, for I understand that our most learned layman, scholar, lawyer and diplomat, Matthew of Irwell, has recently returned from Palestine with an account of some recent discoveries made by the Mohammedans.

◆   Amongst the ancients and the moderns, therefore, there is, and has been, a diversity of instruments with which to study the heights, shadows and motions of the astronomical bodies. Yet the most important instruments of all are our two eyes, for with them we can not only behold the ingenious products of our own earthly ingenuity, but also gaze aloft through the seven planetary spheres to behold the eighth sphere of the stars, which stand before the very Gates of Heaven.

The Astrolabe
by Aelfred
Cliftoniensis,
Emissary of the Archbishop
of York to Rome, with a
preliminary note by the Editors

◆ More news has recently been received of that wonderful instrument called the **astrolabium** or astrolabe. Although its principles are said to have been outlined by Claudius Ptolemy in his **Magna syntaxis**,[7] of which, alas, no copies are known to survive in Europe, an account of a working astrolabe has been brought to Anglia by Brother Aelfred Cliftoniensis, the Emissary of the Archbishop of Eboracum[8] to Rome, who has just returned from a five-year stay in Italy. It is hoped that Brother Aelfred's account will supplement our existing knowledge of the astrolabe that was brought out of Mohammedan Spain some years ago by the Frenchman Gerbert of Aurillac, who has just, in February of this present year AD 999, been elected Pope Sylvester II in place of Pope Gregory V, who died suddenly.

◆ While travelling in the kingdom of Naples, Aelfred met a Mohammedan named Ibrahaim, a merchant of

---

[7] The Almagest.

[8] York.

Damascus visiting Naples on business. Master Ibrahaim had a brass astrolabe, the plate of which was ten thumb-breadths across and was most wonderfully engraved. Its purpose had been explained to him through the good offices of one Levi Ben David, a physician and astronomer of the medical school of Salerno, who was fluent in both the Latin and Mohammedan tongues, as well as in Hebrew and Italian.

◆   Brother Aelfred writes:

The astrolabe is a most ingenious instrument that can be used both to observe the heights and distances of stars and planets, and also to calculate and predict their motions. It consists of several parts. The main part is a heavy brass plate about one-third of an inch thick and ten thumbs (or inches) across. On one side, the plate is hollowed out to about one-quarter of an inch; this is called the mater, "mother" or "womb" of the astrolabe. Its outer raised wall is divided into 360 degrees, with 360 marked at the top. At the top of the astrolabe, at the place called the "crown", is a decorative mount with a brass ring passing through it upon which to suspend the instrument.

◆   Within the mater space are placed five circular plates of thin sheet brass, slightly smaller in diameter than the entire astrolabe. These five plates are engraved on each surface with a total of ten spherical trigonometrical scales. The centres of these brass circular scales - all of which are drilled out with a central hole one-quarter of an inch in diameter - always represent the North Pole, no matter where one is on the surface of the Earth. Upon each of the ten plate surfaces is engraved a stereographic projection of the position of the zenith with relation to the Pole Star, for

The front, or star-plate, of Ibrahaim of Damascus's Astrolabe.
The outside edge of the instrument is equipped with a 360°
scale. Inside is the RETE, or star map (shown shaded). Each
ornamental point represents the position of a star, while its
upper circle is the solar zodiac. This Rete rotates above the
Climate Plate, which carries sky co-ordinates, so that it is
possible to simulate the risings and settings of the stars. (See
pages 61–69 for a more detailed description). Characteristically
for an Islamic astrolabe, it is surmounted by a heavy brass
"crown" through which double suspension rings are passed. The
crown is engraved with the owner's and maker's names, place of
manufacture, and date: A.H.[9] 370 (or A.D. 992). It is also
engraved with Prayers from the Koran.

---

[9] Dates. "A.H." is a Western adaptation of the Arabic word "Hegira" or
Mohammed's flight from Mecca to Medina, on the Red Sea: "Ante Hegira".
In the Christian Calendar, this took place in A.D. 622, and is recognised as
the foundation year of the Muslim religion. Therefore, the Muslim date 370
becomes the Christian date 992 by adding 622.

Ibrahaim of Damascus with astrolabe, stellar globe and
book of tables.

each of ten geographical latitudes. Master Ibrahaim's
astrolabe, designed as it is for use in the Mohammedan
world, has plates engraved for latitudes of Bombay,
Delhi, Mecca, Cairo, Jerusalem and Damascus, but
there is no reason why plates could not also be
calculated and engraved for Rome, Milan, Cluny,
Paris, Prague, Winchester, Lincoln, York, Lindisfarne
and St Andrews in the Christian world. The plates are
called climates,[10] showing as they do the angles at which

---

[10] From klima, Greek for the angle of the Sun's rays.

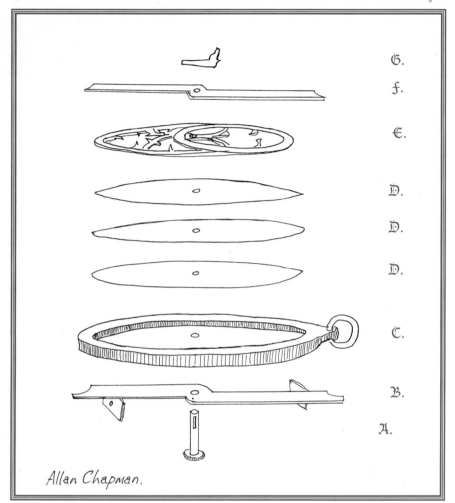

Allan Chapman.

Component parts of the astrolabe. A The brass pin which connects all the other parts. B The alidade, or rule, by which celestial objects are sighted against a scale on the back of the mater. C The mater – the heavy brass body of the astrolabe. D A number of climate plates (in this case three) fit inside the mater. E The rete, or planisphere star map. F The rete rule, for aligning star positions on the rete against the other scales. G The horse, the brass peg that fits through the slot in the brass pin A to lock the whole instrument together.

the Sun's rays strike the Earth at different latitudes, or the heights at which we see the Pole Star.

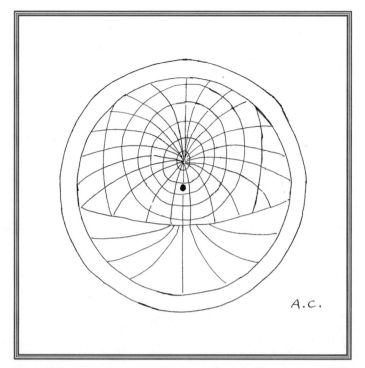

A.C.

One of the astrolabe's climate plates. The central hole corresponds to the North Pole, while the almucantar and equal-azimuth (equal-longitude) lines all focus on the zenith point.

◆ Now, at the centre of each climate plate is a hole, which represents the North Pole, and upon each plate are engraved two sets of lines. These lines are projected stereographically, which is to say drawn from the Earth's South Pole, through a point on the Earth's surface in the northern hemisphere. This surface point is that place or location – let us say the city of York – at which the particular climate plate is intended to be used. As we stand at that point on the Earth's surface, and look directly overhead, then it also acts as our local zenith point: in the case of York, 54° north.

◆ It is this zenith point that provides the key for projecting the two sets of lines that go onto the climate plate. The two sets of lines are stereographically projected in relation to this zenith point. First, there must be a set of concentric circles, drawn between the visible horizon and the zenith, which act as circles of latitude. They are called almucantars, or "arches", in Arabic. Secondly, a set of great circles is drawn through the zenith point, all of them intersecting it. These are lines of "equal azimuth". Now, all of these coordinates are projected upon a sphere, though it requires but a simple act of geometrical construction to drop each of the points and circles on to a flat plate, to obtain a stereographic projection of the sky for a given location.

◆ The closer is a place on the Earth's surface to the North Pole, the nearer are all the convergent lines to the central hole of the climate plate, and hence to the centre of the astrolabe; the reverse is true for places approaching the equator. On Master Ibrahaim's astrolabe, for instance, the plates intended for use in Delhi in India had their zenith projections drawn much closer to the edge of the plate than did the plates intended for use in Damascus. Similarly, on an astrolabe intended for use in Europe, the zenith point on a plate for use in York would be closer to the centre than on one to be used in Rome.

◆ On top of these plates is placed a curiously wrought filigree of brass called, from its appearance, the rete.[11] It carries a circle offset from the centre of the astrolabe by about $23\frac{1}{2}°$, for this circle carries the ecliptic, the Sun's orbit for the year. It is divided into the 12 zodiacal houses, and into 30-digit divisions, so

---

[11] "Net" in Latin.

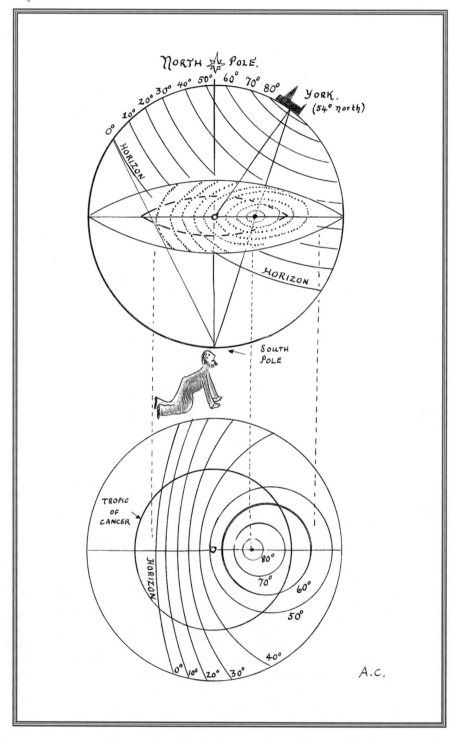

See caption on page 66

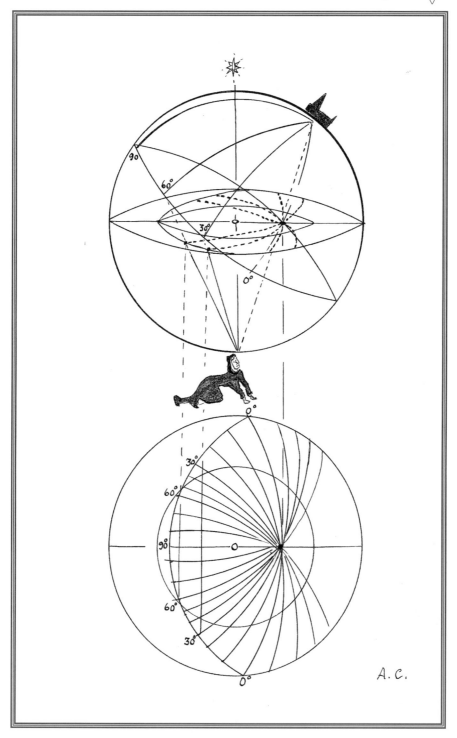

A.C.

See caption on page 66

Pages 64-65: Stereographic projection. Imagine that the Earth is a transparent sphere, with the stars beyond (top left). Imagine an observer who could look up through the South Pole and see the Pole Star directly overhead. The observer next looks at the position of York on the Earth's transparent sphere. He will see York's position offset from the Pole Star, for the Pole Star is overhead, 90° above the equator. The position of York is 54° above the equator.

Now assume that the transparent sphere has a transparent circular plate across its middle, intersecting with the sphere along the line of the equator. We could then "drop" York down onto the plate in such a way that the real position of York on the sphere, its point on the plate, and the observer's eye at the South Pole are all on a straight line. We have now stereographically projected the position of York from the curved surface of the Earth onto a flat plate, without losing its precise geometrical relationship with the position of the North Pole and the horizon.

The next task is to add the lines of equal celestial latitude and longitude. Almucantars (lines of equal latitude) are drawn on the sphere and then dropped down onto the plate (bottom left), and lines of equal azimuth are projected in a similar way (right).

that each digit corresponds to a single day in the year. (Alas, the peoples who lived before the ancient Greeks and who first divided the circle into 360 degrees were not aware that the year is indeed 365¼ days long.) The rest of the rete is made up of a quantity of small brass points, each of which represents an individual bright star. Master Ibrahaim's astrolabe carried the positions of stars best seen from Arabian latitudes, such as Sirius, Aldebaran and Fomalhaut, but for an Anglian astrolabe one might add Capella, or the stars of Ursa

A.C.

The rete of the astrolabe. The central hole corresponds to
the North Pole, and each point represents the position of a
bright star.

Major, and others that are prominent in our northern
skies. All the brass points of the rete form an accurate
stellar planisphere. This planisphere rotates above the
selected climate plate – which is itself locked into
position with a brass lug – so as to simulate the risings
and settings of the stars against the fixed coordinates
of pole, latitude and horizon.

◆ On top of the rete rotates a brass alidade or ruler,
pivoted across the centre of the astrolabe and striking
off straight lines across the rete and the climate plate
beneath. This completes the appearance of the front,
or mater side of the astrolabe. On the reverse side
there is a profusion of finely engraved circular scales
cut into the brass plate, the most important of which is
a calendar or zodiac scale. This scale makes it possible

to place the Sun, for any given date of the year, into
his correct zodiacal house.

◆   What we in Anglia often find so awkward when
attempting to calculate the correct date of Easter Day
in any given year is the Precession of the Equinoxes,
for the First Point of Aries is no longer in the same
position in the sky as it was 1,200 years ago in the
days of Hipparchus. Yet the astrolabe greatly simplifies
the calculation, by enabling us to adjust the date,
ecliptic and zodiac scales so that they correspond to
the present time, and not to the time of the Greeks.
The scales on Master Ibrahaim's astrolabe had been
engraved for the epoch 350 years after the founding of
the Mohammedan religion. When this date is translated
into our Christian calendar, it becomes AD 972. Because
the Precession of the Equinoxes goes so slowly,
however, an instrument adjusted to 972 can be used for
many years to come, without any significant error.

◆   Across the back, or calendar scale, of the astrolabe
is set another alidade, or ruler, pivoted upon its centre,
which is more elaborate than that on the front, because
it carries a pair of brass sights, drilled with holes.
These holes can be used to view a star, or else project
the Sun's image from the front on to the back sight
(thereby avoiding the need to look directly at the Sun),
so that the altitude of any astronomical body can be
measured.

◆   And finally, the entire astrolabe is held together
by a detachable brass rivet, which passes not only
through the central hole of the instrument, but also
through the holes in the five climate plates, the rete,
and the alidades. A socket-hole cut into the shank of
this rivet enables one to insert a delicately wrought

brass wedge, the end of which, in Master Ibrahaim's instrument, was shaped so as to resemble the head of a horse. When the wedge is placed in the rivet's socket, it locks all the parts of the astrolabe together, yet with sufficient slackness to enable the alidades and the rete to be turned independently.

◆ All manner of things can be accomplished with the astrolabe. But first it is necessary to do the following:

(1) Establish your latitude. This is best done by observing the altitude of the Pole Star at night. Take your astrolabe and suspend it by the large brass ring through its crown. Using the alidade on the back scale of the instrument, look through the two sight holes at the Pole Star. In this way you can obtain a direct reading on the 360° scale for your latitude.

(2) Now turn the astrolabe over, pull out the "horse" peg and rivet, and remove the climate plates resting within the mater. Choose the stereographic projection which best suits your measured latitude, and return the five plates to the mater with the correct climate plate uppermost.

(3) Next, return the rete, placing it upon the correct climate plate.

(4) Add the front alidade, or pointer, and put back the peg and horse, to lock the entire instrument back together. You will now find, on turning the rete with your finger, that you have an accurate simulation of the sky for your chosen latitude.

An astrolabe enables one to measure and calculate many things. I have already shown how you may discover your latitude. Yet you may also discover the true hour of day or night.

◆   To find the time by the Sun, you must know whether it is before or after noon for the day in question. (If you are not sure, make several altitude observations of the Sun about 2 minutes apart. If the Sun is getting higher with each observation, it is before noon; if it is sinking with each observation, it is after noon. If the altitude readings are almost the same, it is almost noon.) Then measure the height of the Sun in the sky. From the calendar scales engraved on the back of the astrolabe, you can tell, for your given day of the year, how many degrees the Sun has advanced into its zodiacal sign for that month. Turn over to the mater, and set the front pointer across the ecliptic circle in the rete so that it corresponds to the correct day of the year. Noting the position of the exact day-digit on the ecliptic scale, look through the filigree of the rete, to the climate plate beneath, and carefully turn the rete until that point coincides with the engraved almucantar line that is the same as your latitude.

◆   Then use the central brass pointer to extend this position to the fixed 360° graduated circle that surrounds the mater (which also carries a set of hour divisions), and you will see the exact hour before or after noon when you observed the Sun's position. By using Capella or another bright star, you can obtain the nocturnal hours by a similar method.

◆   Yet many more things can be measured and calculated by this wonderful instrument, such as the lengths of days for given seasons, the motions of the Moon and planets, and even the heights and distances of towers.

◆ I am told that the Mohammedans, like we Christians, need accurate time-keeping for the worship of their God. As we have our daily monastic hours of Matins, Lauds, Sexts, Nones and such, which vary with the long and short days of Summer and Winter, so the Mohammedan has to divide up his day into prayer times. For this purpose, the astrolabe is invaluable. Master Ibrahaim told me that, particularly for a merchant like himself, who must cross oceans and deserts in pursuit of his profession, the astrolabe is a very convenient, quick and portable way of finding his prayer times.

◆ I was also told that Mohammedans and Jews, like we Christians, need to divide up the year for purposes of religious worship. Indeed, when I served the Archbishop of York as Mathematicus et Computator (Mathematician and Calculator), it was my laborious duty to calculate the ever-changing date of Easter Day using the tables of the Venerable Bede of Jarrow. I then discovered from Dr Levi Ben David and Master Ibrahaim that the calculation of both the Jewish Passover and the Mohammedan fast of Ramadan is derived from astronomical criteria based on accurate sightings of the Moon around the time of the Spring Equinox. Although Ramadan can fall in different seasons of the year, the Jewish Passover, like our Christian Easter, always falls in Spring, and is governed by the appearance of the Moon around the time of the Equinox. Christians, Jews and Mohammedans, therefore, must know astronomy if they are to worship God correctly, and in this respect the calendrical function of the astrolabe makes it easier for us to perform our religious liturgies.

◆   Before leaving Naples to return home to Anglia, I paid the sum of forty shillings to one Marcus Orichalchicus Napoliensis[12] to make an exact copy of Master Ibrahaim's astrolabe for me. Dr Levi Ben David has agreed to direct Marcus in the engraving of the climate plates, so that its stars will be more suitable for Anglia, as well as having the Arabic names changed into Latin. He will send it to Anglia by the

Dr Levi Ben David of Salerno guides Marcus Orichalchicus, the brass-worker of Naples, in the engraving and construction of an astrolabe.

---

[12] Literally "Mark the brass-worker of Naples".

R.E.W.C.

Prioress Rachæla teaches the use of the astrolabe to
nuns in Norwich.

Neapolitan agent of Isaac the Banker of Norwich. It
will be delivered into the safe keeping of my
kinswoman, Rachæla Eruditissima,[13] Prioress of the
House of the Consecrated Virgins of Norwich.
Rachæla, who is curious about all learned things, will,
no doubt, wish to have Marcus's astrolabe copied by a
Norwich brass-worker, so that she can teach its usage
to her own nuns, prior to my collection of the original,
when I next visit Norwich on the Archbishop's
business. It is hoped that the instrument will arrive in
Anglia early in the New Millennium.

---

[13] Literally "Rachel the most learned".

Aelfred Cliftoniensis
Written at Milan in Italy on my
journey home, and sent on
ahead to Anglia by fast
courier, whom it is hoped will
cross Switzerland before the
snows block the passes.

7 October 999

A Letter Relating the New and Most Wonderful Discoveries Regarding the Properties of Light,

as made by Alhazen, a physician of Cairo in Egypt. Reported by Matthew of Irwell, Gentleman, Recorder to the Curia Regia, or Royal Council of Anglia, and recently returned from a pilgrimage in the Holy Land

◆ As a Servant of His Majesty Ethelred II, King of England, I received permission over three years ago to

make a pilgrimage to worship at the Holy Places of
Palestine. (Indeed, it was most expedient for me to
leave Anglia at this time for political reasons, as I had
taken a stand against some of His Majesty's bad
advisors, when I urged that we resisted the demands of
certain European overlords and refused to pay the
heavy tax of Danegeld, thereby asserting our historic
independence from foreign control.) His Majesty most
generously gave me letters of safe conduct to carry me
through the territories of sovereigns friendly to Anglia,
as well as diplomatic dispatches to the Emperor Otto
III, who was then in Italy, and to His Holiness Pope
Gregory V. My diplomatic duties completed, I took
ship from Taranto in Italy, and after calling at Malta,
Cyprus, and other islands, arrived at the port of Acre
in Palestine. The Seljuks, who rule in these lands,
treated me civilly, and I visited Jerusalem and prayed
in the Holy Places.

◆   It was in Jerusalem that I met a fellow pilgrim,
who like me was a lawyer and senior civil servant, who
loved books and had a great curiosity regarding all
celestial and natural phenomena. He was Nicholas of
Constantinople, who was Secretary to the Treasury in
the Government of Constantine, Emperor of
Byzantium. Nicholas was a widely travelled and
cultured man who spoke, in addition to his native
Greek, Latin, Arabic and Armenian. His pilgrimage
was also combined with a diplomatic mission, on behalf
of Emperor Constantine, to the Sultan of Egypt,
where he was to negotiate the terms of a trade treaty
concerning the import of Egyptian cotton cloth into
Byzantium. Nicholas, therefore, invited me to
accompany him to Egypt, and to enjoy the protection

of his own diplomatic commission in a part of the world where Christian travellers do not always find a ready welcome.

◆ It was in Cairo that I made the acquaintance of a young doctor who was aged about 35. His name was Alhazen. He was interested in light, and had already made advances that went well beyond those of Aristotle, Ptolemy and other ancient writers. Nicholas, who is himself greatly interested in medicine and astronomy, most graciously acted as my translator, turning Master Alhazen's Arabic into Latin. (For though I had learned some Greek as a student at the Cathedral School in Winchester, I found that my pronunciation was very difficult for Nicholas to understand, and while his Latin was less polished than mine, at least we were soon able to follow each other's Latin pronunciation.)

◆ Alhazen showed me the dissected eye of a freshly dead ox, and pointed out that it contained a circular chamber with a lens made of transparent gristle. He also said that there were organs before the lens to regulate the amount of light that came into it. Although no one knows how God forms the impressions of the outside world in our imagination or understanding, Alhazen showed me that everything at the back of the eye forms itself into a single tough fibre which can be traced to the ox's brain, each eye receiving a slightly different angle of view.

◆ From his experiments on light, moreover, Alhazen feels confident that he can contradict Ptolemy and other ancients who believe that all eyes generate invisible "feelers" that go out before us to intercept the light, to form a kind of invisible touch. Alhazen believes the converse, that light enters straight into the eye without any need for "feelers" to reach out to it. He says it falls upon the lens and produces an image direct! And while it would be exceedingly impious to dissect the body of a dead man or woman (an opinion, indeed, shared by both Christians and Mohammedans), we can none the less cut up the dead bodies of oxen or monkeys as a way of learning how living things work.

◆　In addition to the physiology of vision, Alhazen has investigated many branches of what the ancients call optics and meteorology, to see how light can be bent by air, crystal and water to produce a most bewildering variety of images. Some of the most remarkable are those produced in air.

◆　Alhazen told me that the reason why the Sun and Moon look bigger on the horizon than they do in high heaven is because the air bends or refracts their light. He said that, by studying the first appearance of morning twilight (what is called in the Latin tongue crepusculum), and measuring the exact time that elapses from that first glimmer of dawn to the complete appearance of the Sun on the horizon, he had calculated the depth of the air which surrounds the Earth! For as light travels from the Sun in perfectly straight lines, it first strikes the upper limits of our air

---

Opposite: Alhazen calculated that the first glimmerings of the dawn twilight (or the last at dusk) are visible when the Sun is 19° below the horizon. From this angle, the atmosphere is just able to catch the Sun's rays and refract them down to the surface. (Later astronomers would define "astronomical twilight" as beginning at dawn or ending at dusk when the centre of the Sun's disk is 18° below the horizon.)

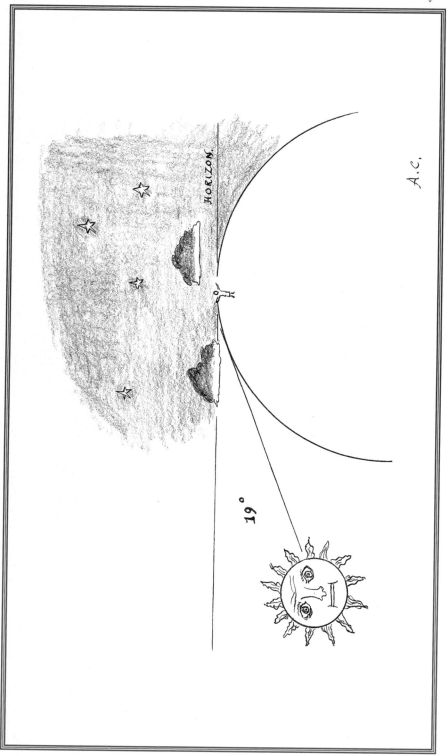

when the Sun himself is still 19° below the horizon.
When these solar rays strike our atmosphere, they are
immediately bent to produce a scattering effect in the
sky. This corruption of the pure white light by our
earthly air, moreover, produces the beautiful colours
that we see at sunrise and sunset, yet which we do not
see when the risen Sun's rays strike us more directly.

◆ Doctor Alhazen shows remarkable ingenuity in
the devising of what might be called "trials" or
"experiments" which attempt to replicate specific
pieces of nature under controlled conditions. I was
especially impressed with his investigation of the
rainbow, which Aristotle wrote about, but failed to
explain satisfactorily. Instead of drawing up
conjectures about the nature of the rainbow, Alhazen
showed me a crystalline or glassy sphere that he had
hung up in a dark room. Admitting a ray of sunlight
through the shuttered window, he made the light enter
the sphere at an angle, upon which it was reflected
twice within the sphere, on its inner surfaces, before
exiting at an acute angle to that at which it went in.
Yet though it was a single white ray of sunlight that
entered the sphere, it was a curious spectrum of
colours that came out of it. No matter how he varied
the angle of the Sun's ray, the order in which the
colours – going from red to blue – emerged from the
glassy sphere was always the same.

◆ This is, indeed, a most ingenious discovery,
showing how the decomposition of white light into
colours in a glass ball reproduces, in miniature, the
formation of the colours of the rainbow when the Sun
shines at a certain angle into a watery cloud. One can
also produce a remarkable array of colours by passing

sunlight through a glass flask filled with water, and the colours thus produced become even more wonderful when one drops a ball of glass into the water beforehand.

◆　Alhazen showed me various other optical devices that he had constructed. There was in his work-room (or more nobly, in the Latin, his elaboratorium) a most curious lathe upon which he could figure rock crystal and glass, to impart precise geometrical curves. He also showed me a mirror of polished metal that was hollowed out, or concave, in the middle. Now, while I have seen such mirrors before and understand that the Romans knew how to figure them, Alhazen showed me a property of these mirrors that is truly amazing. Not only can they concentrate the rays of the Sun to produce a focus, and hence start a fire (an already known property of such "burning glasses"); but he demonstrated that the depth of concavity determines the distance of the point of focus. He told me that each mirror has a fixed point of focus, which is governed by the way in which its curved surface reflects the light. He said he was engaged in trying to develop a mathematical theory, based upon the results of careful experiments, whereby all mirrors and glasses produce images at their foci; this I reckoned to be a most wonderful and illuminating exercise for man's faculty of reason. I also understood him to say that even plain or flat mirrors reflect in a similar precise geometrical way; and while a plain mirror has no focal point, it none the less reflects light in an exact ratio to the angle at which the light strikes the mirror in the first place.

◆　I was, in addition, fascinated to see the images produced by his polished crystals, or lenses. Some were thin slips of figured glass or rock crystal; others were polished cylinders with either flat or curved ends. The cylinders that had flat ends did not distort things when I looked through them, though the ones that were convex – thicker in the middle than at the edges – made things look bigger. These glasses were also capable of producing small, distinct, yet upside-down images of objects upon a whitewashed wall, in addition to coloured fringes. Alhazen told me that, no matter which piece of glass or crystal he used, the coloured fringes always fell in precisely the same sequence, red to blue, and seemed to possess intrinsic geometrical properties.

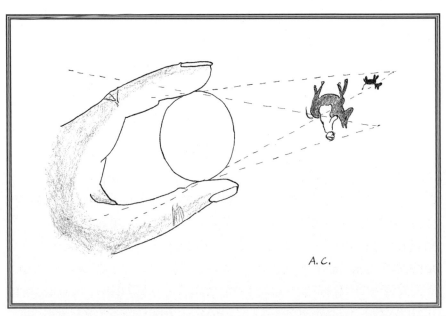

A piece of perfectly clear crystal or glass, ground so that it is thicker in the middle than at the edges, can be used to project an image on to a white screen, once the lens has been focused. An image produced by a lens will be much brighter than one produced by a pinhole. But, like a pinhole image, it will be upside-down, and reversed right to left.

◆ Doctor Alhazen went on to explain to me that, while God did, of course, make the sky blue, he obtained this blueness – as he also obtained the redness of sunset – by using the natural properties of optics, which science he also created. Indeed, Alhazen assured me that he was by no means the first mortal man to realise that the sky's blueness was caused by the laws of optics, for his predecessor, Alkindi, who had lived over 100 years ago, had written on the same. Alhazen was proud of the fact, moreover, that both he and Alkindi were natives of the city of Basra, which, so he said, lies in southern Persia.

◆ When I was examining some of Alhazen's glasses, I found not only that some magnified so powerfully that they made fine silk look like coarse basketwork, but that one of them, when held up to my eye, enabled me to see things more clearly than I had for years. Unlike the others, this glass did not just magnify in a general way, but actually clarified my blurred vision. Doctor Alhazen told me that I was not the first 47-year-old he had met whose naturally weakening eyesight was improved by a correctly shaped piece of glass or crystal. While he had not yet discovered the geometrical rules by which a glass possessing a particular curve was beneficial to the eye of a specific person – for some people, especially younger ones, can see perfectly clearly without it – he said that it was a problem upon which he intended to bestow further attention.

◆ Very generously, Doctor Alhazen presented the eye-correcting glass to me, and as I travelled across Europe home to Anglia, I showed it to several people. Egbert, my own scribe and travelling companion, however, who is about 25, finds that he can see better without the glass, whereas a learned French bishop with whom I stayed on my way home, who is over 70 years old, said that while the glass did indeed make things appear clearer to his old eyes, it none the less failed to make them clear enough!

◆ Yet one of the most amazing things which Alhazen demonstrated to me was how to make Sun-images inside a darkened room. By drawing all the blinds and shutters in his elaboratory, so that the only light which entered did so through a tiny hole in the shutter (creating what is called, in the Latin, a camera obscura), he was able to make a brilliant image of the world outside appear upon the whitewashed wall opposite the small hole. The detail quite took my breath away, as did the bright and natural colours of the noon-tide square outside his elaboratory window, with its moving people, camels and traders, and even birds flying across the sky. Curious as it may seem, though, this perfectly coloured image was inverted and reversed: upside-down and left to right. Alhazen told me that he was still trying to understand how a single tiny hole could produce such scenes, for unlike a glass (which he said should do the same, if only he could make one of the right focal length), there was no corrupting material other than the air through which the light had to pass.

Opposite: Alhazen demonstrates the optics of the pinhole camera obscura to Nicholas of Byzantium and Matthew of Irwell. The street scene outside is being projected onto the wall of Alhazen's darkened elaboratorium.

◆ He assured me that on several occasions he had obtained circular images of the rising or setting Sun by this projection technique. The solar images, however, always seemed curiously egg-shaped on the horizon, due to the distorting power of the air which I mentioned above. Even so, he suggested that one might be able to compute the apparent, or angular diameter of the Sun by this method, if one compared the measured diameter of the projected solar image on the wall with the carefully measured projection distance between the small hole and the wall itself.

◆ Yet what truly amazed me, when I thought of all Alhazen's optical and meteorological wonders, was the order and geometry that underlay them all. For surely, if coloured light is produced when the pure, white celestial light of Heaven is corrupted by contact with the airs and glasses of our fallen and sinful world, how can they follow a geometrical order? For corruption knows no laws of logic, nor passes through such precise and elegant sequences that can be repeated at leisure in an elaboratory. For, as the four sciences of the Quadrivium teach us, geometry is about purity and unchanging perfection, whereas decomposition, being an aspect of chaos, follows no rules. And yet, as Alhazen's demonstrations showed, optics can be expressed geometrically, which only makes me wonder how many other branches of learning might one day be found capable of precise mathematical formulation. Perhaps alchemy and medicine, in addition to optics and astronomy, may be found so.

◆ One day, when I was staying in Cairo, a man who looked and dressed like an Egyptian came to the door of my house. I was surprised, therefore, to discover

that he was a fellow countryman of mine called
Nigellus Anglicanus.[14] Although, indeed, he was not
little, standing nearly six feet high, he did have a
sunburnt complexion, and spoke cultivated, albeit
rusty, Anglo-Saxon and Latin. He told me that he had
travelled to Egypt from Anglia some 25 years ago,
having been inspired by the teachings of our own most
learned Sister Beata of Nunnaminster, and by a vision
of the blessed St Santha of the Holy Well of Ffestiniog,
to give up worldly concerns, and live as a hermit in the
desert of Egypt, after the fashion of the Christian
hermits of 600 or 700 years ago. He said that, in his
desert wanderings, he had travelled so far to the south
that at mid-day, in June, he stood in his own shadow,
the Sun being directly overhead. He also told me of a
deep well down which the Sun shines on mid Summer
day, to be reflected in the water at the bottom. Now,
since even on mid-Summer day the Sun always casts a
shadow in Cairo, I can only assume that Nigellus
Anglicanus had ventured as far south as the Tropic of
Cancer, $23\frac{1}{2}°$ above the equator, of which Ptolemy
speaks. The true existence of such a Tropic, and the
absence of shadows thereupon in June, must provide a
clear physical proof that the Earth is indeed a sphere.

◆ Nigellus told me that, though he lives as a hermit
in isolation for months at a time, he occasionally meets
other men. Many years ago he had learned to speak
Arabic, and it was from a recent chance encounter that
he heard from a passing camel-drover that an English
Christian was currently in Cairo under the diplomatic
protection of the Sultan, which encouraged him to seek

---

[14] Literally "Nigel, (or the little dark man) of England".

me out. Nigellus also mentioned that the contacts with isolated travellers over the years had led him to believe that the peoples of India possessed a wonderful knowledge of the movements of the Heavens and a system of mathematical notation vastly superior to that used either by the Arabs or by us Christians of Europe. Unfortunately, he knew no precise details of Indian astronomical knowledge.

◆   One thing of great wonder which I did see during my time in Cairo was the Pyramid of Pharaoh, King of Egypt. Several ancient writers in both the Greek and Latin tongues speak of it, but no written account can prepare a visitor for what he will see, for in size it is bigger than any church in Christendom. Doctor Alhazen, who kindly took me to see the Great Pyramid, which stands only just outside Cairo, says that it is perfectly aligned, north, south, east and west, and contains a shaft that points in the direction of the North Pole. I found this quite surprising, for we in Christian Europe have always believed that the Pyramid was hollow, having been built by the Jews to serve as a granary for the Pharaoh, some time after Joseph foretold the years of famine by interpreting the Pharaoh's dream. Yet this story of its being a granary cannot be true, for the Pyramid is quite solid, like a mountain of granite aligned with the Heavens.

◆   After spending some six months living in Cairo, in a house enjoying the diplomatic immunity conferred upon Nicholas, I set sail for home. As Autumn was beginning to set in, knowing of the violent winds that blow in the Eastern Mediterranean from the account of St Paul's shipwreck in *Acta apostolica*, I was thankful to arrive safely at Brindisium in Italy. I called to pay my respects to the Pope once again, only to find that Gregory V had died suddenly in February this year, 999, and that the former Archbishop of Ravenna, Gerbert, was now Pope Sylvester II. I am honoured to carry papal and diplomatic dispatches from this most universally learned man back to King Ethelred and the Archbishop of Canterbury in Anglia.

◆ After a rough crossing of the English Channel from Francia, during which time my small ship was blown well off course by easterly gales so that we all feared for our lives, we thanked God for our safe landfall on a beach at Torre Bay in Devon. The Abbot of Torre most generously gave me rest and hospitality, and it is from the lodgings which he kindly placed at my disposal that I dispatch this account of my travels and meeting with Doctor Alhazen, in the hope that it will not be too late for inclusion in the Yearbook.

> Cum multa salute[15]
> Matthew of Irwell
> Torre Abbey, Devonshire
> ante diem XV Kalendas Ianuarias M[16]

---

[15] "With many greetings".

[16] 18 December AD 999.

ook Review

Book of the Constellations of the Fixed Stars, by Abd Al-Rahman Al-Sufi. First issued in Arabic, but now available in Latin

◆ There is no doubt that Al-Sufi ranks as one of the most outstanding astronomers of modern times. He died in 986, though news of his demise did not become generally known until quite recently. He compiled books on astrology and on instruments such as astrolabes, and he carefully listed the names of the stars (most of which are of course Arabic). His catalogue, reviewed here, is regarded as his most important contribution, at least to astronomy. Most of it was completed by the year 964.

◆ It is based on the famous catalogue by Ptolemæus (Claudius Ptolemy), which was itself loosely based upon that of Hipparchus. However, Al-Sufi's new catalogue is an improvement on its predecessors. The positions, magnitudes and colours of all the brightest stars are given; there are two drawings of each constellation, one as seen from the inside of a globe and the other from the outside (the figures on the charts in this Yearbook are based on Al-Sufi's).

◆   Of special importance are the magnitudes of the stars, but there are some discrepancies here; for example, the star Zibal, in the constellation of Eridanus, is rated by him as of the 4th magnitude, whereas Ptolemy made it of the 3rd. The present reviewer has confirmed that the magnitude is in fact 4. Is it possible that this star – Zibal – varies in brightness? No variable stars are known, and the generally held view is that the Heavens are perfect, permanent and unchanging. Indeed, it might be considered heretical to think otherwise. But the discrepancy is still puzzling.

◆   There is one improvement which could surely be made, and which I am bold enough to suggest. Catalogues, even Al-Sufi's, describe the stars by their positions in the firmament. There could no doubt be some similar method – such as numbering the stars in each constellation; thus in Orion Betelgeux could become 1 Orionis, Rigel would become 2 Orionis, and so on. Names are very useful, but there are at least 2,500 stars to be seen and remembered, and to find 2,500 suitable names is not easy. Numbering would be better, and for this purpose the new notation (1, 2, 3, ...) is clearly more convenient than the old (I, II, III, ...).

◆   But for the moment, this catalogue by Al-Sufi remains the best available, and I have no doubt that it must surely remain the standard for centuries to come.

M. L.

Obituary

Abul Wafa

Many and distinguished are the astronomers of the Arab region. Mistaken though some of their beliefs may be, yet they earn our respect. One of these who must be remembered is Abul Wafa. We have heard news of this death, in the year 998, at the age of 59.

◆ Alas, we do not yet appreciate the extent of his achievements. He produced an Almagest (not, of course, to be confused with the great Almagest of Ptolemæus), he developed some of the instruments described in this Yearbook, and he paid close attention to the motions of our nearest neighbour in the heavens, the Moon.

◆ A full appreciation of his life and work will, we hope, be included in our Yearbook for 1001.

C. J.

Astronomy of the Next Millennium

◆ As we come to the year 1000, and look forward to the new Millennium, we must ask ourselves: What developments are in store, both in astronomy and in astrology?

◆ Perhaps the astrological outlook is the more predictable of the two; we know now how the heavenly bodies move, and how they influence our lives. There are seven nearby bodies (the Sun, the Moon and the five planets) and there can be no more, since seven is the mystic number. Refinements may be made, but the principles of astrology have been well and truly said, to our lasting benefit and to the glory of God, whose divine judgement is written in the stars and is conveyed to us.

◆ Astronomy is not so predictable. Our great deficiency is that we cannot see clearly enough. We look at the Moon, and we can make out its lands and its seas; Plutarch called it "earthy", with mountains and ravines. But can there be some way in which we could see it as it would be if we were closer to it? Can we develop eyes which will bring distant objects closer? It was said that Lynceus, steersman of the Argo, had eyes which could see further than those of any other man; can we not only emulate Lynceus, but even

95

surpass him? This is something we do not yet know, but perhaps, before 2000, we will make new "eyes" which will show us the lands and seas not only of the Moon, but also of the planets.

◆   Then, can we solve the riddle of the universe itself? How big is it, and how old is it? We see the spheres of the planets, the Sun and the Moon, and we see the sphere of the fixed stars, all of whom must assuredly be at the same distance from us; but how high is the sphere of the stars, and what lies beyond – if indeed there is a beyond? Science of today cannot tell us. It may be that the science of 2000 will bring us some of the answers, but we cannot be sure. It may be that God Himself has made sure that no matter how far we probe, we must remain in ignorance.

◆   And finally, and of the greatest importance of all, is there a future for ourselves reaching out to 2000? It has been said that the world will end before the close of the present Millennium. Well, we have only months of the present Millennium left - and if our readers are able to peruse these words, now being written by your humble scribe, we will know that at least one dire prediction has been falsified.

◆   Who knows? The coming Millennium may produce scientists as great as Thales and Ptolemy, philosophers as great as Aristotle, observers as great as Al-Sufi, and statesmen and leaders as great as King Alfred. Time will tell. This the last Yearbook of our era; in twelve months we welcome the year of grace 1001 - and with it, all our hopes for the future.

C. L.

Appendix

A Note About the Conversion Table (p. xi)

You can demonstrate the superiority of the Arabic numbering system simply by attempting long division in Roman. Take, for example, the sum of 506 gold pieces and try to divide it equally between 22 men using Roman notation.

A History of Times to Come?

Publisher's Note: It is far from clear where this appendix came from. In places it seems almost to be looking back on the Yearbook, as if from some distant future year. In this respect it is most strange. It must be said that, although the writer clearly knows something of the contributors to the Yearbook, phrases like "of questionable historical authenticity" are hardly calculated to endear the writer to those who have laboured hard and long to produce this Yearbook!

◆ This Appendix should therefore be regarded as probably entirely fanciful, most likely placed with the text by some mischievous Brother at Winchester with too much time on his hands. Nevertheless, we include it here as our gesture to the frivolity and celebrations that usher in the New Millennium.

A Note About the Authors and Certain Persons in the Text

The monks Aelfred Cliftoniensis, Albert Pendleburiensis, and Allanus Salfordiensis, and the diplomat Matthew of Irwell, are individuals of questionable historical authenticity. Likewise, some of the persons with whom they had dealings, most particularly Ibrahaim the Merchant of Damascus, Dr Levi Ben David of Salerno, Gregory the Archimandrite of Constantinople, Marcus Orichalchicus, Nicholas, Treasurer of Byzantium, Nigellus Anglicanus the hermit, Aethelwulf Contusissimus, and the lay brethren Simon Stultus and Daniel Cloacinus, seem to have left little verifiable trace on the historical record. The same, alas, is also true of Lilia the pie-maker of Oxenforde, the learned Sister Beata of Nunnaminster, Rachaela the Prioress of the House of Consecrated Virgins of Norwich, Helena the Cantrix of Paris, and St Santha of the Holy Well of Ffestiniog.

◆ Yet what Aelfred, Albert, Allanus and Matthew actually did, their travels to foreign lands and their interests in exotic branches of learning, are none the less entirely consistent with the behaviour of a handful of learned men in their professions around that time. A limited number of Christian pilgrims were also allowed to visit the Muslim-controlled Holy Land. Southern Italy, moreover, was frequented by sophisticated Muslim merchants like Ibrahaim. Europe's first medical school was founded at Salerno near Naples around AD 850, and there were Jewish physicians like Levi Ben David who knew Hebrew, Latin and Arabic, and who taught there. Italy was also a centre of fine craftsmanship, with men like Marcus Orichalchicus practising in its major cities.

◆ All others whom they met, wrote of or read, have firm and uncontested places in the historical record. Gerbert of Aurillac (c. 945–1003), moreover, whom the Emperor Otto III put forward as Pope Sylvester II in 999, after previously being Archbishop of Ravenna, was one of the most dynamic scholars of the age; he is said to have travelled to Muslim Spain, he certainly taught at Rheims, and he took an active interest in astronomy and science. He was also the first Frenchman to become Pope. Similarly, Alhazen (965–1039) was the most distinguished experimental physicist to emerge out of medieval Islam, and his researches into optics, ophthalmology and meteorology were profoundly influential, as were those of his predecessor Alkindi.